After Earth Day:

Continuing the Conservation Effort

After Earth Day:

Continuing the Conservation Effort

Max Oelschlaeger,
Editor

University of North Texas Press

First Edition 1992

95 94 93 92   4 3 2 1

Requests for permission to reproduce material from this work should be sent to the
University of North Texas Press, P.O. Box 13856, Denton, Texas 76203-3856

This book is printed on Envirotext, Acid Free paper, which meets EPA Recycled
Standards.

The paper in this book meets the minimum requirements of the American National
Standard for Permanence of paper for Printed Library Materials, Z39.48—1984.

Library of Congress Cataloging-in-Publication Data

After Earth Day : continuing the conservation effort / Max Oelschlaeger, editor.
        p.   cm.
    Includes bibliographical references and index.
    ISBN 0-929398-40-8
    1. Environmental policy—United States—Congresses. 2. Conservation of
natural resources—United States—Congresses. 3. Human ecology—United
States—congresses. I. Oelschlaeger, Max.
HC110.E5A633   1992
333.7'0973—dc20   92-922
                                                      CIP

CONTENTS

*
**

Introduction

Max Oelschlaeger

THE TWENTY YEARS between Earth Day I and Earth Day XX are
not, as history goes, a long stretch of time. Yet we learned a
fundamentally important lesson over that short period: there are
no shortcuts from ecological disequilibrium to sustainability
(leaving those controversial notions undefined). The pertinent
facts are frightening, and they speak for themselves, since
whatever the environmental achievements in the two decades
between Earth Days I and XX—and I am not demeaning them—
on the whole conditions are worse. The world's population
continues to explode. Climate heating or the so-called greenhouse effect is no longer a scientific hypothesis but a global
reality that promises an uncertain future. The Amazon rain
forest is undergoing relentless assault. Thousands, if not tens of
thousands, of plant and animal species have been and are being
driven into extinction. *Science*, perhaps the most prestigious
journal of science in the United States, states unabashedly that
humankind is endangering all species, including itself.[1] Consequently, one is tempted to advance the conjecture that two
decades of environmental activism indicate, whatever our limited
successes, nothing more clearly than an inability to stem the
basic drift of Western culture toward ecological breakdown.

The skeptic, of course, might respond that we are only now
able to measure the "true dimensions" of the ecological crisis.
Hence, whatever the insufficiencies of the past twenty years,
tomorrow will bring a greener future since we now know—
measurements in hand—how to blueprint a sustainable world.
Yet the skeptic's argument founders on those aforementioned

facts, the realities of life on earth. Reread, for example, *The Environmental Handbook*, published originally in 1970, a few months before the first Earth Day (April 22, 1970). The book became a virtual bible for the environmental teach-ins and consciousness raising activities associated with Earth Day I, going through three printings between January and April, 1970. What startles the contemporary reader is that virtually all the problems that characterize our present ecocrisis are identified there. Paul Ehrlich writing on "Too Many People." David Brower on the wilderness and its preservation. Kenneth Boulding on the paradox of an economics that defines success in a way sure to undercut nature's economy. Lynn White on the culpability and responsibility of Judeo-Christianity. Gary Snyder (in an unsigned essay) on the necessity of creating an old-new mythos that would lead us into a postmodern age. And the list goes on and on.

The upshot, then, given that the actual problems have been widely known at least since Earth Day I *and* that they continue to worsen, is that Earth Day is a misnomer. Any authentic celebration of the earth inherently involves more than a twenty-four hour media blitz or a week long conference. Conservation must become a reality in our daily lives, in the basic institutions and practices of our culture, if we are to find our way to sustainability.[2] Should we celebrate an Earth Millennium instead?

Which brings us to the point of this collection of essays, originally delivered at a conference held at the University of North Texas during the week of Earth Day XXI. *After Earth Day: Continuing the Conservation Effort* celebrates and renews the spirit of Earth Day I through XX, not as media spectacles and mass consciousness raising efforts, but as exemplifying the sustained commitments of many different people and organizations to a common cultural effort. We have learned over these last two decades that there are no quick fixes for the environmental crisis, that conservation is not one but many things, that more than anything else conservation involves lots of different people doing diverse things that may eventuate in a greener future. The evolution of our advanced industrial culture from its present condition of disequilibrium toward a condition of sustainability depends, more than anything else, upon the continuing efforts of everyone. Conservation is people in business, in the university, in science and technology, and everyone else who cares about the earth acting locally but thinking globally.

The essays are gathered into five sections, basically reflecting the actual sessions held over the four days (April 15–18, 1991) of the conference itself: Conservation Politics; Environmental Science Today and Tomorrow; Conservation, Economics, and the Corporate Effort; Environmental Philosophy; and Religion and Conservation. The collection itself serves as the inaugural volume in a new Philosophy and Ecology series from the University of North Texas Press and also provides an enduring public record of the conference. Together, the essays have many interesting convergences and divergences, and readers will want to read essays in different sections in the context of others (for example, Nieswiadomy in contrast to Paehlke). Ideally the reader can gain insight into conservation over the last twenty years and some perspective on where it might be going during the 1990s.

While vigorous discussion at the end of each presentation was the rule rather than the exception, limitations of space preclude publishing the dialogue. (The sessions were well attended, averaging nearly 200 persons.) Yet a common theme emerged from the conversation. Americans are very concerned about the relations between our socioeconomic processes and the natural world which underpins and sustains our culture. And increasingly they realize that there is no one to help us but ourselves. Sustainability hinges on the unrelenting efforts of all of us working together, contributing in a wide variety of ways to a sane, green future.

✻✻✻

Section one, Conservation Politics, gathers three insightful essays written by George Sessions, America's best known deep ecologist, Robert Paehlke, whose book *Environmentalism and the Future of Progressive Politics* is a near definitive study of environmental politics, and Pete A. Y. Gunter, one of Texas' leading conservationists who has battled for thirty years to preserve the Big Thicket. Placing these essays at the beginning reflects the reality that any collective movement toward a sustainable society will be effected through presently existing political institutions. Together they sweep the spectrum of political possibilities, Sessions sketching the promises of a green politics or radical environmentalism that would transform our polity, Paehlke dealing with the potential for environmentalism, defined as a diverse but increasingly potent political movement, affecting change through the two-party system, and Gunter narrating the autobiographical and practical realities of environmental activism.

Paehlke's essay, "Environmental Politics and Policy: The Second Wave," gives us reason to think that the "second wave" environmental movement might accomplish some if not all of its objectives through democratic processes, partly because environmentalists are better organized and more powerful than during the "first wave" of environmental activism, and partly because the second wave has learned much from the failures of the first. Today the public increasingly recognizes that there are no quick fixes, that transformations in attitudes and values are fundamental, and that alterations in how we work, live, transport, feed and house ourselves are in the offing. But such changes will not necessarily be onerous, involving radical changes in life style, nor draconian, involving the emergence of "administrative despotism," if we begin to effect timely modifications now through the existing system, including market-based policy tools such as user fees and subsidy removal. Paehlke does not believe that we have time to reinvent our political-economy, and provides examples of solutions (that are neither left nor right) worked out through presently existing local, regional, national and international means.

Sessions' essay, "Radical Environmentalism in the 90s," develops an important distinction between two kinds of environmentalism: *anthropocentric environmentalism* oriented largely toward urban problems, especially pollution, and *ecocentric environmentalism* that considers conservation as involving the entirety of Earth/Gaia rather than just narrowly conceived human interests. (Paehlke's paper touches on this distinction, but Sessions develops it in some detail.) While radical environmentalism has flourished in the intellectual community, Sessions believes that it has largely been ignored within the mainstream conservation movement—dominated by anthropocentrism. Such an anthropocentric fixation has been shown by conservation biology to be perilous since human beings are part of the natural cycles of life on earth: to ignore

Gaia's health is to ultimately undercut our own. Sessions also presents a sketch of the history of Earth First!, an organization that began out of frustration with the actions (or inactions) of mainline anthropocentric conservation groups, and the attack made by social ecologists on Earth First! specifically and deep ecology more generally. In the concluding section, Sessions turns to his vision of the future of radical environmentalism.

Gunter's essay, "The Perils of Conservationist Politics: Life in the Trenches," reads as a wry commentary on more than thirty years of hands on experience as an environmental activist, beginning with a letter he wrote as a Yale undergraduate to Senator Ralph Yarborough concerning the Big Thicket and ending with the creation of the Big Thicket National Preserve some two decades later (1974). But Gunter's essay subtly communicates far more than a personal history. Namely, the reality that individuals with a good cause can collectively join and make a difference even in the face of overwhelming opposition, and that the individuals best equipped to do this are local people with their roots set in real places. Such people will not emerge unscarred from the political trenches, but the system does sometimes work. And the work goes on, unceasingly, for those who commit themselves politically. As I write (September, 1991) Pete is off to Washington, D.C., once again presenting testimony to the United States House of Representatives on the Big Thicket.

❋❋❋

Section two, Environmental Science Today and Tomorrow, contains essays by Curt Meine, whose *Aldo Leopold: His Life and Times*, is the definitive Leopold biography, Ken Daugherty, one of the nation's leading researchers on Refuse Derived Fuels (RDF'S),[3] Neil Evernden, whose books bear thoughtfully on the issue of defining ecology itself, and Ken Dickson, a restoration ecologist and Director of the Institute of Applied Science at the University of North Texas. Clearly, no solutions for ecocrisis can be conceptualized outside the framework of scientific research and knowledge. Just as clearly, scientific knowledge alone cannot ensure success, partly because of the enormous scope and diversity of contemporary science, partly because of the difficulties inherent in applying knowledge to practical problems, and partly because science inherently involves a host of philosophical and ethical issues that preclude easy passage from knowledge of means to determination of ends to be achieved. The diverse array of essays in section two sweep across this entire spectrum, from Daugherty's research on refuse derived fuels to Meine's account of the sudden appearance of conservation biology, from Dickson's vision of a truly ecological education to Evernden's argument for an ecology of limits.

Meine's essay, "Conservation Biology and Sustainable Societies: A Historical Perspective," details the almost revolutionary emergence between Earth Days I & XX of a new interdisciplinary science that may do in the twentieth century what physics did in the seventeenth: that is, transform the way we understand ourselves in relation to nature. Conservation biology arose in part in response to the continued drift of conservation in the last two decades, especially in terms of the failure to preserve life on earth, that is, protect biodiversity. Similarly, the new terms "sustainability" or "sustainable societies" grew out of the scientific recognition that *development*, at least as conceptualized within the prevailing Western socioeconomic paradigm, was antithetical to biodiversity. Meine's essay is leavened throughout with a wealth of historical detail, including many pertinent observations on and reflections of the life and work of Aldo Leopold. There can be no doubt that *conservation biology* has arrived as a legitimate science, as confirmed in a recent issue of *Science*, dedicated to "Perspectives on Biodiversity."

Daugherty's presentation, co-authored with Cheryl L. Brooks, "Municipal Solid Waste to Fuel," based on a research project that he has directed over the last decade, shows how science helps to solve environmental problems that are not as dramatic as, say, holes in the ozone, but nonetheless real. One such problem is the enormous mass of waste generated by American families and industry, garbage that threatens our soils and ground waters with contamination, and refuse that consumes prodigious amounts of space for landfills. Daugherty's research group has found that RDF'S can provide ecologically safe and economically efficient energy. Furthermore, by processing waste for RDF'S virtually all recyclable materials are reclaimed. Metaphorically, this work represents a contemporary kind of alchemy, where something precious, energy and a enormously reduced volume of waste materials, is derived from a base material—garbage that would otherwise be buried in landfills.

Evernden's essay, "Ecology in Conservation and Conversation," is a philosophical reconsideration of the science of ecology itself. If Meine helps us to discern that the evolution of science has yielded conservation biology, and Daugherty provides a practical use for science, then Evernden helps us conceptualize what tomorrow's ecology might be. He contends that ecology, caught up in the business of the description of nature, remains oblivious to the sociocultural environment that enframes it. In consequence, today's ecology, embedded in a cultural tradition that unconsciously assumes human development as the highest good, is at best a foreshadowing of what tomorrow's might be: a science that recognizes and accepts nature's way, that is, the reality of limits on all species. Evernden's essay is more than anything else a candid and penetrating assessment by a scientist of the ecological enterprise. He argues that science does not exist in an objective but a social space, and that a science which fails to recognize this fact is a science that will ultimately fail to serve society.

Ken Dickson's contribution, co-authored with Andy Schoolmaster and Sam Atkinson, "Educating Environmental Scientists for the 21st Century," begins with the recognition that the need for environmental education is acute, but that the inherent interdisciplinarity of the field poses challenges for educators. No one field, Dickson argues, provides an adequate base for the investigation of environmental problems, climate heating being a case in point. Drawing on his experiences as an environmental educator designing new graduate level programs at UNT, Dickson discusses the goals, program design, curriculum, teaching methods, and the obstacles to achieving interdisciplinary teaching. The ultimate aim is to produce a generation of environmental scientists trained to think proactively and holistically while also possessing the requisite skills of analysis and measurement.

❋❋❋

Section three, Conservation, Resource Economics, and the Corporation, contains essays by Jenny Cheek, Project Manager for Distribution, Mary Kay Cosmetics (Dallas), E. E. Spitler, a recently retired executive for the Chevron Corporation (San Francisco), and Michael Nieswiadomy, Director of the Center for Environmental Economic Studies and Research at the University of North Texas. While environmentalists often indict marketplace economics as the primary cause of the blight upon nature, there is no reason to think that any overnight economic revolution will change the basic ways of industrialized societies. Corporations did not create the culture they serve. As Pogo said, we've met the enemy, and it is us. So viewed, we can begin to see the modern corporation in a different light. Mary Kay Cosmetics, Inc., did not create the demand for cosmetics any more than Chevron invented the hydrocarbon economy. More generally, not even the most ardent conservationists live as ecological saints since there are few alternatives to the prevailing system of transportation, agriculture, housing and so on. These essays presume that our political-economy will not be reinvented from the ground up, but more likely transformed by a process of disjointed incrementalism as it comes to grips with ecocrisis. Clearly, these essays remain, as deep ecologists and others could point out, anthropocentric. Just as clearly, they sketch out a path and take a few steps toward an ecologically improved, if not ideal and indefinitely sustainable, society.

Cheek's essay, "The Corporate Responsibility to the Environment," gives an overview of recycling at Mary Kay Cosmetics, Inc., a leader in the city of Dallas for ecologically affirmative corporate action. Through an in-house program, voluntarily initiated by the company, Cheek's corporation has taken strides toward the goal of reducing its production of waste (thus interdicting the cycle addressed in Daugherty's RDF research above). Voluntary efforts by corporations to act in ecologically responsible ways are often difficult in a competitive environment; but recycling is one area where all corporations can likely take action. Cheek also discusses her involvement with the Corporate Recycling Council of Dallas, an innovative program that to date has offered two seminars on recycling attended by more than 400 corporate executives, including representatives from major metropolitan areas around the country interested in starting similar programs. A third seminar, featuring speakers on product life-cycles, was held during September, 1991.

Spitler's essay, "The Energy Business and Conservation," draws on his long and interesting career with Chevron, one of the major players in the energy industry. He poses tough questions for conservationists about alternatives to the hydrocarbon economy, the standard of living made possible by utilization of fossil fuels, and the willingness of consumers to change their energy related behaviors. Spitler also surveys a variety of issues from the perspective of the energy industry, including government regulation (legislation is often desirable, and criminal rather than financial penalties are more efficacious in influencing corporate behavior), the corporate "insiders" culture (a new generation of ecology-minded executives is beginning to make its presence felt), global warming (reduced demand through legislation and more efficient consumption are realistic short-term alternatives), and responsibility to future generations (since hydrocarbons are irreplaceable resources). More than anything else, Spitler argues for a proactive rather than reactive corporate involvement, taking action that expands the purview of executive decision-making beyond the bottom line to include environmental ethics.

Nieswiadomy's essay, "Economics and Resource Conservation," attempts to show how resource economics can help a market society come to grips with ecocrisis. There are many philosophical issues that Nieswiadomy does not consider, such as Nicholas-Georgescu Roegen's critique of marketplace economics from the standpoint of the entropy law, Donald McCloskey's examination of the rhetoric of economics, and Mark Sagoff's argument that all social choice cannot be reduced to economic decision-making. Yet read as a claim that economic analysis can shed at least some light on a dark debate about cultural choice and resource allocation, Nieswiadomy's case is credible. As conservationists have discovered, the most politically efficacious arguments against environmentally hazardous or destructive practices are often economic. Nieswiadomy argues, for example, that the policies of the United States Forest Service are not only an ecological but also economic disaster. More controversial is his argument that the way to protect at least some endangered species, such as the African elephant, is through the market mechanism. By privatizing elephants, he claims, individuals acquire an interest in protecting their long-term interests which necessarily include those of the species.

✳✳✳

Section four, Environmental Philosophy, contains essays by Dolores LaChapelle, Director of the Way of the Mountain Center (Silverton, CO), E. C. Hargrove, editor of the leading journal in the field of environmental philosophy, *Environmental Ethics*, and Chair of the Department of Philosophy and Religion Studies at the University of North Texas, and Michael Zimmerman, a deep ecologist and Chair of the Department of Philosophy at Tulane (New Orleans). Collectively, these essays indicate the diversity of environmental philosophy. LaChapelle's essay is rooted in the environs of the San Juans, reflecting her deep and abiding interaction with the land community with which she identifies. Hargrove's paper reflects the more technical kinds of arguments that characterize the field of professional environmental ethics (although his essay ends on an ironic note). Zimmerman's piece stands, appropriately, in a middle zone between subjective immediacy and scholarly objectivity, and perhaps offers a *via media* between the two.

LaChapelle's essay, "Deep Ecology and Nonhuman Others," begins with some observations about where we were on the occasion of Earth Day I and where we seem to be going now. She sees only one way in which the destruction of life on earth might cease: by rediscovering the importance of people living in place, re-establishing natural relations sustained over the years and even generations with a land community. As the conservation movement stands, she argues, abstract theories (the *Gaia* hypothesis, ecosystem ecology, resource economics) deceive people into thinking that they know nature when they do not. LaChapelle illustrates from her own experience of living in southwestern Colorado, among the flora and fauna, the San Juan mountains themselves, and the town of Silverton. Her paper concludes with examples of how humans can live with the land community in sustainable, nonintrusive fashions.

Hargrove's essay, "Weak Anthropocentric Intrinsic Value," is a rigorously argued treatment of environmental rights and value theories. Such theories came of age shortly after Earth Day I, leading to the origin of the journal *Environmental Ethics* and the publication of dozens of books on the subject in the last decade or so. Hargrove examines several different theories, including the positions of Paul Taylor, Holmes Rolston, III, Baird Callicott, and Bryan Norton. Ironically, since the paper is intensely philosophical, readers will discover that Hargrove argues for the legitimacy of everyday judgments made by "real" people. These individuals, who are not concerned with the technical arguments of philosophers, offer what Hargrove terms "weak anthropocentric intrinsic value" arguments for the conservation of nature. Ordinary language arguments—for example, that a cave should be preserved because it is beautiful or that a wilderness area should be protected because it is irreplaceable— work, according to Hargrove, not because they are underpinned by epistemological and metaphysical theories, but rather because they express our individual desires articulated in the context of our cultural traditions and contemporary problems.

Zimmerman's essay, "The Future of Ecology," begins with the widely known distinction between deep and shallow ecology, but then departs from convention by asking if deep ecology itself does not remain embedded in a narrative tradition that assumes a progressive human future. He contends that even deep ecologists (radical environmentalists, ecological activists) want to take the future of the world in their own hands, a mission exemplified by the notion of global ecology. Zimmerman next examines the positions of individuals who have either denied the possibility of anything beyond acting locally (e.g., Wendell Berry) or deconstructed deep ecology through the methods of postmodern textual criticism (e.g., Jim Cheney). The essay then asks if there is a *via media*, an Aristotelian middle. The future of ecology, Zimmerman concludes, involves the attempt to recognize the otherness of nature and the difference of non-Western cultures as well as the self who can identify with Gaia and the reality of global ecological processes.

✳✳✳

Section five, Religion and Conservation, presents essays by Susan Bratton, a widely published ecologist who also writes on Christian environmental ethics, Elinor Gadon, author of the best-selling *The Once and Future Goddess*, and Max Oelschlaeger, editor of this volume, who is completing a book on the role of religion in a time of ecocrisis. These essays span an array of contemporary thought on religion and environmental crisis and largely converge on a common center. Each author affirms the central idea that apart from religious traditions and discourse there is little hope for a sustainable and socially just society. Bratton's approach to a "new" Christian ecology is simultaneously ecological, since Christians need more specific knowledge about the realities of life on earth, and historical, since many possibilities for a Christian environmental ethic can be found in history. Gadon's paper offers a challenge to Bratton's Christianity by finding new possibilities in a contemporary neopaganism that reaffirms the Goddess wisdom of the Paleolithic and Neolithic. Oelschlaeger's essay identifies a vital role that religion and only religion can play in coming to grips with ecocrisis.

Gadon's essay, "Metaphors for Birthing: Towards a New Creation Story for the Age of Ecology," argues that the Judeo-Christian creation story—most pointedly, the metaphysical separation of the divine from the secular, and humankind from nature—is the historic root of ecocrisis, including the modern endeavor to understand nature scientifically and to dominate it technologically. The Goddess traditions and mythologies of antiquity offer, in her opinion, a viable alternative, a new creation story for an age of ecology. She articulates a cosmogony of the generating earth, illustrating with examples from a variety of non-Western cultural traditions. Gadon also thoughtfully draws out the mythic implications of contemporary ecofeminist art, such as that of Judy Chicago, Ann Mendieta, Cristina Biaggi, Mimi Lobell, Vijali Hamilton, and Judith Anderson— women who dare to explore the psychic and symbolic connections between Mother earth and their own bodies.

Bratton's essay, "The 'New' Christian Ecology," bypasses the conventional critiques of Judeo-Christianity's role in ecocrisis by asserting that more than anything else some new directions in Christian ecotheology are in order. She argues that—viewed in terms of the actual relations to and treatment of the earth—the problem *is not* necessarily orthodox Christian cosmology but the lack of concrete strategies for action. She sketches, for at least a broad middle spectrum of Christian believers, the outlines of a positive environmental ethic consistent with the orthodox Bible story (e.g., Genesis 1, where God affirms the inherent value or intrinsic worth of all the creation). Bratton also finds exemplary Christian individuals and communities, such as the desert fathers and mothers, the Celtic monastics, and the Franciscans, who once lived in harmony with the earth and nonhuman others, and who may now serve as models for the development of a contemporaneous Christian environmental ethic.

Oelschlaeger's essay, "Caring for Creation: Religion in a Time of Ecological Crisis," argues from a sociolinguistic perspective that religion is our last, best chance to achieve sustainable forms of social existence since god-talk more than any other kind of discourse legitimates the human project. Every religious tradition, including most importantly (at least in the context of North America) the Judeo-Christian tradition, and virtually every denomination within that tradition, has either already articulated or can articulate a compelling rationale to treat nature with respect. (Read in terms of Hargrove's essay, god-talk is the primary source of weak anthropocentric intrinsic value, that is, those descriptions of and deliberations about natural processes and natural entities that either fall outside the bounds of instrumental value or trump claims of instrumental value.) In sum, Oelschlaeger's thesis inverts the conventional wisdom that religion is the cause of environmental crisis.

Special thanks are due Tom Preston, Dean of the College of Arts and Sciences, and Fran Vick, director of the Press, for their support, Ike Orloff for coordinating the conference, the student groups and individuals who contributed resources and energy, the financial sponsors, and the many participants whose oral presentations and personalities made the conference a success and whose essays made this volume possible.

NOTES

1. See *Science*, August 16, 1991.

2. Arguments abound over the basic meaning of "conservation." The essays by Neil Evernden, Curt Meine, George Sessions, and Robert Paehlke below are astute treatments of the issue. Resourcism, or the progressive utilitarian definition of conservation which emphasizes efficiency and utility, largely dominates public consciousness. But rival definitions, such as John Muir's (in the late 1800s) and Michael Soulé's, challenge the anthropocentric bias of resourcism, although they remain at this date a "minority report."

3. Cheryl L. Brooks, a member of Daugherty's RDF research group, wrote the essay collected here, based on Daugherty's presentation at the After Earth Day Conference, April 17, 1991. Correspondence is to be directed to Daugherty, University of North Texas, Department of Chemistry, Denton, TX 76203.

*
**

PART ONE

CONSERVATION POLITICS

Environmental Politics and Policy: The Second Wave

Robert Paehlke

IT HAS BEEN twenty-one years since the first Earth Day was celebrated. With hindsight we can see that it was celebrated at the crest of a first wave of contemporary environmental concern. That first wave had begun to subside by the mid-1970s. By the period of economic stagflation (1979-1982) environmentalism in North America was in considerable political retreat. However, as economic stability was re-established, a second wave of environmental concern, stronger than the first, surged forward. By Earth Day 1990 we were again at the crest of a wave. It would appear that, in the case of environmental issues, there is not a simple one-time process of public concern, governmental action, and declining interest.[1]

As well, the two recent surges of environmental concern were preceded by the longer history of the conservation movement. The environmental movement can, however, be distinguished from the conservation movement in several important ways.[2] The environmental movement has been relatively more oriented to urban issues such as pollution and energy sustainability than to concerns regarding wilderness and forests. The environmental movement has perhaps exhibited deeper doubts about economic growth and technological "progress." Accordingly, the environmental movement has at times raised more fundamental ideological, political, and institutional questions. Conservation organizations, however, have continued their important work in the contemporary era and operate in parallel with the newer environmental organizations. Some leading organizations with conservationist origins, such as the Sierra Club, have addressed both conservation and environmental issues, though their emphasis has generally remained on the former.

It is most interesting to consider some of the ways in which the second (current) wave of environmentalism differs from the first in terms of politics and policy. First wave environmentalism arose largely as a set of "motherhood and apple pie" issues. At first, there was minimal political resistance in public. Significant political resistance first arose in the mid-1970s regarding a perceived conflict between environmental protection and many of the possible prospects for resolving the post-1973 energy crisis. Environmentalists opposed off-shore drilling, open-pit coal mining, nuclear power and dams on wild rivers. This politically difficult position was significantly improved by the "soft path" arguments of Amory Lovins and others. Nonetheless, environmentalism was soon seemingly overwhelmed by the deregulatory mood of the late 1970s.

The origins of the second wave of environmentalism can be seen in the immediate reaction to the arrival of Ronald Reagan in the White House. Both environmental and conservation organizations increased their memberships rapidly throughout the 1980s, even as public attention turned elsewhere and the administration tried to systematically roll back the environmental progress of the 1970s. The membership of many established organizations doubled during the 1980s at the same time that new organizations such as Greenpeace came from nowhere to claim memberships in the millions.[3] Even in politically difficult times, when the public was justifiably alarmed about inflation and unemployment, the organizational strength of environmentalism was advancing. After the mid-1980s, when the economy was more stable, the second wave had a base from which to build.

Impetus for the second wave has also been rooted in a particular set of issues. Since 1986 public attention has been riveted by global warming, holes in the ozone layer, a continent-wide crisis in solid waste management, the rapid loss of the critical habitat provided by tropical rainforests and old growth forests generally, and several dramatic incidents such as the Exxon Valdez oil spill. These issues and events have helped to establish more firmly the global character of the environmental crisis. They have also placed energy efficiency and materials use into a central position within environmentalism. Finally, the protection of wilderness has re-emerged as a centrally important issue. In the second wave of environmentalism, conservation and environmental concerns are more fully integrated than they were in the 1970s.

Contemporary environmentalism, then, combines the traditional concerns of conservation—loss of wildlife habitat, overexploitation of forests, endangered species, water management and allocation, and quality recreational space—with the newer, more urban-oriented health and sustainability concerns of the environmental movement. In the early 1980s the pollution concerns broadened from the air and water quality emphasis during the first wave to include hazardous wastes and climatological concerns. The second wave has also seen a shift in concern from particular pollution problems to global effects and to more generalized problems such as indoor air quality (related to cigarette smoke and other pervasive hazards) as well as the household hazardous wastes present in every landfill. As well, environmentalists are now emphasizing waste reduction and thus a focus centered more on the core patterns of industrial production and consumption.

All of these changes provide grounds for optimism that the second wave of environmentalism will have significant and lasting effects. Nonetheless, it is still the case that the overwhelming majority of North Americans are yet to appreciate fully the implications of global environmental problems. Most do not, for example, see their concern through into their voting habits.[4] But more than that, only a limited number of people appreciate that environmental concerns provide a fundamental challenge to the structure (though not necessarily the market-based organization) of our economy, our political institutions, our policy-making habits, our cultural values, and, in a phrase, the North American way of life. Global warming and the garbage crisis have, however, caused more and more people to see that environmental concerns are not short term problems with easy, technical, add-on solutions. They are, indeed, political and economic and cultural and social problems requiring change at every level: individual, community, corporate, institutional and governmental.

One specific example of the type of changes that are necessary may be useful here. The amelioration of global warming will probably require a significant reduction in the production of carbon dioxide within the developed economies, especially in North America. In turn this need will likely require a reduction in the number of automobiles in North America and/or in the number of miles which they travel. A recent study of forty large cities around the globe indicates that North American urban residents drive twice as many miles as do Europeans of comparable income levels.[5] As well, some North American cities—most notably Houston, Phoenix and Los Angeles—are significantly more auto-intensive than are New York, Chicago, or Toronto.

On a worldwide basis the determinant of public transportation use, average auto trip distance, *and* the proportion of those cycling or walking to work is urban density. That is, residents of Paris and Amsterdam drive on average half as many miles because their cities are more compact. Those cities are not, however, more crowded or more congested. They are arguably more, not less, pleasant and liveable.

Thus, when one considers the non-sustainability of fossil fuel supplies, air and water quality, oil spills, and the greenhouse effect in combination one recognizes that the configuration and density of most North American urban settings are inappropriate and unworkable. The contemporary North American urban form is a mistake which can be corrected through the infilling of medium-density housing, including town houses, low-rise apartments, and condominiums. New urban patterns may well emphasize tighter simpler, grid patterns. Changes of this sort imply corresponding shifts in values, culture, industrial structure, and public policy at all levels. This example is typical of the contemporary agenda of environmental politics and policy. End-of-the-pipe solutions are no longer adequate.

During the first wave of environmentalism terms like ecology, epidemiology, and toxicology established a place within every day discourse. During the first wave the policy tools of choice were regulations, especially emissions standards. Conceptually, the second wave has centered more often on achieving sustainability and global change.[6] The tools of choice include multipartite treaties such as the Montreal Protocol, mandated product and process changes such as mandatory recycled content in newsprint, and investments by electrical utilities in energy efficiency as a source of supply. As well, the return-to-the land individualism of the 1970s has given way to politically street-smart environmental organizations.

The new politics of environmentalism is not consistently left, or right, or center. Environmentalists themselves are widely divergent in their views on "non-environmental" issues. Interestingly, as we will see below, they are thus free to select from the full range of possible political allies and policy tools. The balance of this chapter will take a closer look at both policy and politics, beginning with the former.

Regulatory and Non-Regulatory Approaches to Environmental Policy

IT HAD BECOME current by 1990 to argue in favor of market-based approaches to environmental protection.[7] Such approaches appeal to politically conservative and politically moderate environmentalists and to some non-environmentalists who would be just as happy with minimal environmental protection. Many environmentalists prefer a mixed instrument (regulatory and non-regulatory) approach. Other so-called market-based approaches are essentially deregulatory. That is, the former approach considers environmental regulation to be necessary, but not sufficient. The latter, however, seeks to *replace* a regulatory approach with market-based techniques. The difference can be critical. Regulatory sticks are necessary to encourage an enthusiastic acceptance of non-regulatory carrots. As well, an *exclusive* dependence on market-based tools can create significant political difficulties (some of which are discussed below).

Caveats aside, a wide variety of market-based policy tools can be adapted to an environmental protection role. Many, though not all, market-based tools are traditionally advocated by the political right. However, their explicit adaptation to environmental protection is quite new and potentially of interest across the political spectrum. Three traditional tools of the right that are important here are: (1) user fees, (2) subsidy removal, and (3) deficit reduction.

Environmental user fees are now very much a part of the second wave of environmentalism. Carbon taxes are widely advocated as a means of limiting the production of greenhouse gases and promoting both energy efficiency and fuel substitution. In effect, users are charged for the share of the burden they place upon the global climate. President Bush's acid precipitation package also included proportionate charges for the production of sulfur dioxide by electrical utilities. But perhaps the most direct of user fees are those now utilized in solid waste collection by several municipalities including Seattle. These cities charge for domestic solid waste pick-up by the volume of waste generated.[8] Sorted recyclables are picked up at no charge and thus there are clear economic incentives to reduce waste and to recycle. Other environmental user fees include: (1) individual metering of electricity use in apartments or office complexes, (2) charges for parking at workplaces to encourage car pooling, cycling, walking and the use of public transit,[9] (3) metering for domestic water use, and (4) time-of-day electricity and freeway tolls to reduce the need for greater peak-load capacity.

The removal of direct and indirect governmental subsidies can also provide a considerable increment of environmental protection. For example, large water and electricity users are frequently subsidized by either small users or taxpayers. The effect is a considerable disincentive to conservation expenditures and the promotion of resource uses which may be both economically and environmentally inappropriate. All subsidies to energy production are, by this logic, a mistake. This would include the considerable taxpayer support which continues to exist, for example, for the nuclear industry. But perhaps most interesting here are recent studies which calculate the governmental subsidies to automobile ownership. Subsidies in this case include: (1) highway construction and maintenance, (2) police (40% of whose efforts are auto-related), (3) snow and ice removal, (4) the tax deductibility of providing parking for employees, and (5) traffic planning and management. One recent figure estimates the total subsidy in the U.S. at $2,400 per vehicle per year.[10] Such practices actively discourage the use of public transit and all non-automotive means of transportation. Now-standard land-use planning practices, in combination with this subsidy regime, virtually assure urban sprawl and high levels of air pollution in North America's cities.

As well, consistently high levels of governmental debt may also be environmentally, as well as economically, problematic. Debt, both public and private, is a means of achieving present consumption without requiring present payment. To the extent that these expenditures have attendant environmental impacts deficit spending helps to generate damage which might otherwise not occur. But more than this, these practices impose on future generations both environmental damage and a limited fiscal capacity to remediate that damage. Future generations are injured doubly. On this point fiscal conservatives and environmentalists can concur.[11]

By way of a sharp contrast, effective environmental policies can also be achieved through measures more traditional to the political left. Barry Commoner, for example, argued quite effectively that the resolution of the energy crisis of the 1970s (not, one assumes, the last such crisis) required nothing less than the nationalization of the oil industry.[12] Other analysts have, of course, argued that private utilities perform better environmentally than do public utilities.[13] This is a complex matter, but a governmentally-owned energy industry if subjected to direct participatory public controls is environmentally an interesting option.[14] Questions of ownership aside, it is also arguable that environmental protection can be achieved through the use of more stringent regulatory enforcement including the use of criminal penalties.[15] Such an approach (traditional within hard-nosed populist and left-of-center regimes) would, of course, also provide a considerable competitive advantage to those corporations which have already undertaken sound environmental practices.

There are also a number of innovative environmental policy tools which are not easily associated with either a left, or a right, perspective. These would include governmental procurement, altered utility investment practices, and new varieties of persuasion and dispute resolution. This list of approaches is also more common within the second wave of environmentalism. Regarding procurement, many state and local governments now purchase recycled paper and other environmentally advantageous products. As well, they are also, increasingly, encouraging newspaper publishers to use paper with a significant proportion of recycled content. But more and more jurisdictions are also coming to appreciate the breadth of the environmental potential of governmental procurement.[16] Governments purchase almost everything: buildings, vehicles, chemicals, food, and cleaning supplies. They could purchase low-impact (or fewer) chemicals, organically grown food, and more energy-efficient buildings. They could subsidize walking, cycling, and public transit for governmental employees. They could develop "naturalized" urban parks. The second wave of environmentalism frequently demands better environmental performance from each and every public institution including hospitals, school boards, universities, and municipal governments. The "simple" (federal or state) regulation of private corporations is more often now seen as but one necessary policy step among many.

Several electrical utilities, especially in the New England states, have recently adopted important new practices. In 1988, the Conservation Law Foundation, a Boston-based environmental group, won a new type of ruling before the Connecticut Public Utilities Commission.[17] Utilities in New England must now invest directly in energy efficiency in new building design, in commercial and domestic retrofits, and in altered industrial processes using less electricity. Some even supply efficient appliances and lighting to homeowners. Energy efficiency is far and away their least cost source of "new" electric supply. The change is that those efficiency improvements are now treated as if they were a supply source; these investments are included in the utilities' rate base. Utility profit levels are thereby protected while demand growth is reduced. Since every other new source of supply (other than efficiency improvements) would have significant environmental impacts, the environmental protection achieved in this manner is considerable.

Lastly, there is now wide enthusiasm within Canadian governmental and business circles for what are infelicitously called *multi-stakeholder consultative processes*. The attempt here is to bring together government, industry, labor, environmental organizations, and community groups. The objective is consensus regarding a wide variety of economic and environmental practices, policies, and projects. Businesses and governments, it is assumed, wish to improve their environmental performance. Environmentalists, it is assumed, will accept the competitive climate and insufficient revenues as grounds for limited and/or delayed action. All parties agree to focus on the positive steps that can be taken now. Obviously, the advantage within this type of process falls to those who are comfortable with the status quo—failure to reach mutual agreement results in no change from present practice. Nonetheless, there have recently been some gains for environmentalists from face-to-face contacts. These have been achieved within a wider climate of strong public environmental concern. In such a context, all parties prefer to take, or at least to appear to take, positive steps towards more effective environmental protection.[18]

Thus, the age has passed when pollution abatement regulations and standards were "the" (singular) means of achieving environmental protection. Many tools are available, each with a particular political configuration arrayed on either side. It is arguable that there is a good deal to be gained in utilizing a diversity of policy tools. Each approach will spread the economic burdens and the economic gains in a somewhat different pattern. Environmentally, the positive effects associated with a multiplicity of tools may well be cumulative. Practices will be changed and habits undone by a combination of economic incentives and persuasive inducements, backed up by feared penalties. Politically, of course, each mix of policy tools will mobilize and appease a differing mix of economically interested groups and organizations. Many environmentally sound policies fail to clear this hurdle. It is, thus, to the political, as opposed to the policy, realm to which I now turn.

Environmental Politics at the Crest of a Wave

THE FIRST EARTH DAY (1970) was preceded and followed by a variety of legislative initiatives almost all of which created an attendant administrative process at the federal and/or state level. Consider, for example, the National Environmental Policy Act (NEPA), the Resource Conservation and Recovery Act (RCRA), the Toxic Substances Control Act (TSCA) and Superfund legislation. The second wave of environmental concern seems to have set in motion a process which is more socio-economic than administrative. It might be argued that where governmental regulation and enforcement were typical of the 1970s, the 1990s will emphasize altered patterns of consumer purchases (green products) and individual waste management habits (recycling, composting, household hazardous wastes, etc.). This is not to say that either approach (administrative or socio-economic) is the more desirable. Indeed both are necessary to the achievement of adequate environmental protection.

There are at least two reasons for this conclusion. First, the changed behaviors associated with second wave initiatives frequently address different dimensions of environmental problems. Using recycled paper saves overcutting of limited (and ecologically essential) forests. But a de-inking (paper recycling) plant may or may not be itself a significant source of pollution. This in turn may well depend on water pollution regulations and enforcement. Thus both approaches are necessary. The progress based in the second wave is only truly effective in association with that based in the first. Second, protection based wholly in market-induced change may generate considerable political resistance. There are some very good reasons for this.

Individuals and families with lower income levels spend, for example, a much higher proportion of their more limited funds on energy. Simply raising the price of energy (through taxes or other means) thus hits the poor proportionately harder. Further, the poor frequently have a lesser capacity to adopt alternative behaviors. The elderly and disabled are less able to cycle and walk, working single parents are much more dependent on some "convenience" items, and all of the poor are less able to pay for insulation upgrades or newer, more efficient, appliances and vehicles. And, needless to say, the less-well-off are much more vulnerable to job displacement. Even when new jobs replace old jobs the act of changing jobs (and family location) can be very difficult without a considerable savings cushion.

These are just some of the realities which lie behind the fact that environmental concern has not been seen through into positive results in referenda and electoral politics. Even where the feared costs of environmental protection are not real many persons are vulnerable to the distortions and half-truths of counter campaigns. There remains, for example, a considerable gap between perception and reality as regards the employment impacts of environmental protection. There are, in fact, more jobs—not fewer—associated with returnable beverage containers, recycling, pollution abatement (in most instances), energy efficiency (versus new energy supply), and the increased use of public transit.[19] The principle political problem here is that those existing jobs which are threatened by change are defended. The new, alternative employment opportunities have no incumbents and therefore no defenders. Even where there are economically interested advocates they are frequently less significant to the economy and more dispersed (e.g., local diaper services as against a few very large international corporations).

Environmental politics in the 1990s will succeed to the extent that environmentalists (1) communicate that apparent employment/environment conflicts are frequently a false dichotomy, and (2) design policies that appeal both to the less advantaged *and* to those whose first concerns are environmental. These are sometimes the same people, but frequently not. Even though the poor more frequently bear the brunt of pollution impacts, they cannot always make those impacts their first concern. More than that, of course, the poor in the United States are frequently only marginally involved politically as regards any issue. The most important challenge of the 1990s may be the design and selling of policies which protect the environment and at the same time mobilize the less politically active segments of society. Recent events in the province of Ontario and elsewhere would suggest that the combination of environmental and (moderately) progressive politics may be a formidable combination in the present political climate.[20]

Some Policies of Mutual Appeal for the 1990s

IN CONCLUSION, I offer a short list of policy initiatives which have potential appeal to both environmentalists and those groups which have traditionally supported progressive politics (urban dwellers, trade unionists, minorities, and the poor). While there are many possible policies with such an appeal, those offered here are put forward as examples. It is important to appeal to this broad constituency to avoid in the future the anti-environmental political backlash which dominated politics and policy in the early 1980s. The following policies would seem appropriate to the task:

1. Establish urban core revitalization programs with an emphasis on medium density, affordable housing and improved public transportation. More compact cities are more environmentally sound. They can result in air quality improvements and significantly higher levels of energy efficiency. Such a policy is also labor intensive and would have very positive social consequences,

depending on design quality and such other factors as the equity of housing and employment opportunities.

2. Provide energy efficiency upgrades for the aged and for other lower income persons and families. One of the best approaches to this policy would see electrical utilities leasing high efficiency lighting fixtures and appliances. As well, heating energy suppliers could be transformed into suppliers of least cost heating and cooling energy. Several municipalities and states have taken effective initiatives in this policy area.

3. Policy initiatives could be found which address the subsidy now provided to automobile owners. However, even though many poorer (especially older and younger) urban residents do not own automobiles the simple removal of subsidies could economically cripple other poor persons, especially where quality public transportation is not available. More appropriate solutions would include increased taxes on large new cars combined with rebates on fuel efficient cars. Alternatively, a partial rebate from general revenues could be paid annually to automobile non-owners in the form of transit passes. Most simply the subsidies to public transit and to automobile owners could be equalized over time.

4. Improvements in medical, educational and social services should be seen as having, as well, positive environmental impacts. These are employment intensive expenditures with only very minimal environmental impacts per job. The same is true of public expenditures on the arts.

5. On the other hand, there are many governmental expenditures which have negative environmental impacts and few positive social impacts. Reduced expenditures could be applied to some of the options above or could be used to reduce the tax burden on the less advantaged. Expenditure reductions might include the down-sizing of government-owned vehicles. As well, government employees could be urged to use public transportation rather than receiving mileage allowances for private cars. More significant in dollar terms, water, energy and resource supply subsidies could be eliminated. In addition, municipal parks and the landscaping around government buildings could be "naturalized" (less lawn-oriented) resulting in lower maintenance costs, and reduced energy and chemical use. Needless to say, military expenditures also frequently have unintended environmental impacts and could be reduced in the present climate.

Conclusion

IN THE TIME that has elapsed between Earth Day I and Earth Day XX we have learned a great deal about environmental politics and policy. We must learn a great deal more, especially about the interconnections between the two. The second wave of environmentalism has made clear that there are many policy instruments that are important to the achievement of environmental protection. The fact that there are waves—rising and falling trends in environmental concern—suggests that environmental organizations should look for ways to broaden their political base when the wave is at its crest.

NOTES

1. See Anthony Downs, "Up and Down with Ecology: The 'Issue-Attention Cycle,'" *Public Interest* 28 (1972), 38–50.

2. A more detailed distinction is set out in Robert Paehlke, *Environmentalism and the Future of Progressive Politics* (New Haven: Yale University Press, 1989), esp. at 21–22.

3. See James A. Tober, *Wildlife and the Public Interest* (New York: Praeger, 1989).

4. Riley E. Dunlap, "Polls, Pollution and Politics: Public Opinion on the Environment in the Reagan Era," *Environment* 29 (July/August, 1987), 6–11, 32-37.

5. Peter Newman and Jeffrey Kenworthy, *Cities and Automobile Dependence: An International Sourcebook* (Hants, England: Gower Publishing, 1989).

6. I use sustainability rather than sustainable development advisedly. Sustainable development, as articulated by the World Commission on Environment and Development in *Our Common Future* (New York: Oxford U.P., 1987), may well be an oxymoron.

7. See, for example, Robert N. Stavins, "Harnessing Market Forces to Protect the Environment," *Environment* 31 (January/February, 1989), 4–7, 28-35.

8. Seattle Solid Waste Utility, *New Garbage Collection Services and Rates* (Seattle: Seattle Engineering Department, 1988) and *On the Road to Recovery* (Seattle: Mayor's Office, 1989).

9. See Richard W. Willson and Donald C. Shoup, "Parking Subsidies and Travel Choices: Assessing the Evidence," *Transportation* 17 (1990), 141–57.

10. Reported in, for example, Francesca Lyman, "Rethinking our Transportation Future," *E Magazine* 1 (September-October, 1990), 34–41.

11. See the discussion on this point in Paehlke, *Environmentalism and the Future of Progressive Politics*.

12. Barry Commoner, *The Poverty of Power* (New York: Bantam Books, 1977).

13. James Q. Wilson and L. Richardson, "Public Ownership vs. Energy Conservation: A Paradox of Utility Regulation," *Regulation* 9 (September/October, 1985), 1–17, 336–38.

14. For a longer discussion see Robert Paehlke, "Government Regulating Itself: A Canadian-American Comparison," *Administration and Society* 22 (February, 1991), 424–50.

15. See Russell Mokhiber, "Crime in the Suites," *Greenpeace* 14 (September/October, 1989), 14–17.

16. See Robert Paehlke, "Toward a New Environmental Protection Strategy: A Survey of Options with Emphasis on Governmental Procurement," *Business in the Contemporary World* 2 (Summer, 1990), 30–37.

17. Christopher Flavin, "Yankee Utilities Learn to Love Efficiency, *Worldwatch* (March/April, 1990), 5–6.

18. See Michael Howlett, "The Round Table Experience: Representation and Legitimacy in Environmental Policy-Making," *Queen's Quarterly* 97 (Winter 1990), 580-601 and the chapter by Douglas Amy in Robert Paehlke and Douglas Torgerson, eds., *Managing Leviathan: Environmental Politics and the Administrative State* (Peterborough, Ontario: Broadview Press, 1990).

19. Again, for a more extended discussion see Paehlke, *Environmentalism and the Future of Progressive Politics*.

20. In 1990 Ontario elected the New Democratic Party (NDP) to a majority government position for the first time in history. The NDP has combined traditional democratic socialism with a strong environmentalist agenda.

Radical Environmentalism in the 90s

George Sessions

What is radical environmentalism?[1]

ANY ADEQUATE CHARACTERIZATION of radical environmentalism must include both its theoretical and practical aspects. Radical environmentalism thus involves both *theoria* and *praxis*. The theoretical groundwork was laid in the 19th century with the naturalists Thoreau and Muir, and continued into the 20th century with literary figures such as D. H. Lawrence, Aldous Huxley, Robinson Jeffers, Gary Snyder, and Edward Abbey, and with biologist/ecologists such as Aldo Leopold, Rachel Carson, Paul Ehrlich, Paul Shepard, and Michael Soulé. Beginning in the 1960s, philosophers, historians, and other academics such as Lynn White, Theodore Roszak, Roderick Nash, Arne Naess, Paul Taylor, Baird Callicott, and Warwick Fox entered the radical ecological dialogue. Academic radical environmentalism continues in the pages of such journals as *Environmental Ethics*, the *Trumpeter*, and in other periodicals and books.[2]

Radical environmental activism has been influenced and guided by this academic debate over philosophical basics. Contemporary radical environmental activism has been exemplified in groups around the world attempting to live bioregionally, and in direct action groups such as John Seed's Australian rain forest actions, in Greenpeace, the Sea Sheperds, Earth First!, and in some actions by animal rights groups.

In both theory and practice, the philosophical basis of radical environmentalism is the concept of ecocentrism, which continues to be inspired by recent insights from the science of ecology. Viewed negatively, ecocentrism is a rejection of anthropocentrism in all its forms. Viewed positively, it is an affirmation of the idea that all the wild species of the planet have an equal right, along with humans, to exist and flourish largely without interference by humans in their natural habitats; in this respect no species is privileged. Radical ecological consciousness, then, is ecocentric consciousness together with ecological wisdom (ecosophy). If carried to a deeper philosophical level, there is a spiritual/religious component to ecocentrism involving a psychological and emotional identification with nonhuman individuals, species, and ecosystems resulting in what Arne Naess calls an "ecological self." And beginning with Thoreau and Muir, there has also generally been a tendency toward a pantheistic identification of God with Nature. As Max Oelschlaeger clearly demonstrates in his remarkable reinterpretation of Western culture (*The Idea of Wilderness*), radical environmental consciousness involves a major departure from traditional and modern Western modes of thought. This search for a postmodern cultural paradigm has its roots in Darwin, Thoreau, Muir, and Aldo Leopold and leads to the recultivation of Paleolithic consciousness.[3]

One of the key insights of ecological consciousness is the realization that humans (and human activity) have taken over far more than their ecological share of the planet, resulting in devastating ecological damage to the Earth such as habitat loss for the other species and the present accelerating rates of species extinction. The major priorities of radical environmental activism, given the present rates of destruction of wild Nature and the resultant danger to the integrity of Gaia, have been, and should be, to redress this imbalance by (1) promoting the protection of the remaining wild and semi-wild habitat on Earth; (2) promoting the restoration of wild Nature to protect wild species and the ongoing natural evolutionary processes; (3) promoting the stabilization and reduction of the human population on the planet; and (4) promoting ecologically benign ways of life for humans.

Recent "risk assessment analysis" by scientists has borne out the priorities of radical environmentalism and the deep long-range ecology movement. They have found that the most serious threats to the integrity of the biosphere are the exponential growth of the human population, the greenhouse effect, ozone layer depletion, and species extinction as a result of habitat loss. Various forms of urban pollution and pesticide poisoning, usually focused upon by the Environmental Protection Agency (EPA), the general public, "social ecologists" and many anthropocentric reform environmental groups, have a much lower risk factor.

Population biologists have argued that 1 to 2 billion people living lightly on the planet would be sustainable given the ecological requirements of maintaining a carrying capacity for all species. A human population decrease from its present level to that level (by humane means such as steady low birth rates) would also be good for humans and for the diversity of human cultures, as well as for wild species and ecosystems. Arne Naess once suggested that a population of that size might achieve an ideal balance on the planet consistent with the ecocentric goals of radical environmentalism. Such a balance would consist of 1/3 wilderness (wild species habitat), 1/3 "free nature" (where there are mixed communities of human and wild species living in largely non-domesticated ecosystems), and 1/3 cities, roads, agriculture, etc. for intensive human inhabitation of the planet.[4]

While various forms of non-violent direct action have been characteristic of modern radical environmental activism, I have here purposely characterized radical environmentalism more in terms of philosophical beliefs, consciousness, lifestyles, and long-range ecological goals designed to promote a sustainable ecological balance for all the inhabitants of the Earth.

The history and development of radical environmentalism

RADICAL ENVIRONMENTALISM has a history which begins about the middle of the 19th century. Thoreau and Muir were the early founders and both were essentially ecocentric. As if to under-score his commitment to ecological equality, John Muir once claimed that if a war were to break out between humans and bears, he would tend to side with the bears.[5]

Ecocentric environmentalism became more clearly defined after World War II, as the insights of the science of ecology began to influence environmental thinking. In the late 40s Aldo Leopold's "land ethic" promoted the health of the planet's ecosystems as ethically primary and was thus essentially a statement of "earth first" (an emphasis upon the ecological health and integrity of the Earth's ecosystems and wild species). Also in the late 40s concerns were raised from an ecological standpoint over the exponential growth of the human population by William Vogt, Sir Julian Huxley, and others. In the 1950s, a concern for halting human population growth was integrated into the ecological perspective vigorously promoted by Aldous Huxley.[6]

In the 1960s, radical/ecocentric environmentalism rapidly gathered steam beginning with Rachel Carson's *Silent Spring*, which led to the tremendous outpouring of public support for environmentalism on Earth Day 1970. During this period David Brower was attempting to radicalize and ecologize the Sierra Club. And virtually all of the biologists who were popularizing the science of ecology for the public, from Raymond Dasmann to Paul Ehrlich, were calling for a halt to human population growth.

New environmental concerns over urban pollution entered the arena as a major part of the environmental agenda beginning with the 1960s. Typically, those spokesmen whose main concern is urban environmentalism have tended to be philosophical anthropocentrists, such as Barry Commoner and Ralph Nader. For example, Barry Commoner (with his Marxist "social justice" background) claimed in the early 70s (and still does) that there is no human overpopulation problem. Stephen Fox (in his history of environmentalism) points out that these newer "man-centered environmentalists" of the late 60s (such as Commoner and Nader) distrusted the wilderness protection concerns of the Muir-Brower tradition, were primarily urban and pollution oriented, and had little or no interest in wild Nature or the protection of wild ecosystems and wild species habitats.[7] Urban pollution problems are clearly a crucial part of the environmental crisis, and have now resulted in such staggering global problems as massive air and water pollution, the greenhouse effect, ozone layer depletion, and acid rain. But there are wider ecocentric issues, mentioned above, concerning the long-range ecological integrity of the Earth's ecosystems which a fixation on narrowly anthropocentric urban pollution problems does not address.

As a result of their narrow focus, environmentalists whose concerns are directed primarily to anthropocentric urban pollution problems fail to grasp the full dimensions of the ecological crisis and its philosophical implications. Ecocentrism and the ecological perspective is a radical post-modern philosophical departure from the development of anthropocentric modernist Western thought (which sees the non-human world essentially as "resources" for humans) and, as a result, it has been difficult to sustain in the crisis atmosphere of the environmental developments of the 70s and 80s. David Brower was ousted from the Sierra Club in 1969 partly for his radical views and actions, and mainstream reform environmentalism (both the traditional wilderness and the newer urban/pollution wings) has essentially reverted to secular anthropocentrism, huge bureaucratic structures, "yuppie" materialistic values, and pragmatism and traditional interest-group politics. Designated wilderness areas are again seen by the new yuppie Sierra Clubbers, for example, as areas of superlative scenery and playgrounds for "industrial tourism," rather than primarily as sanctuaries for wild ecosystems and unmanaged habitat for wild species, as John Muir (the founder of the Sierra Club) and the ecologists have viewed them.[8]

By the late 70s environmentalism (both radical and reform) was all but eclipsed as Americans turned to conservative anti-environmental Republican leadership. But the accelerating global environmental crisis could not be ignored forever, and this eventually led to a new public outpouring of concern resulting in Earth Day 1990. Many overriding ecological issues, however, were not emphasized, such as the need for human population stabilization and reduction, the need to protect wild ecosystems and species habitat, and the crucial need for industrial countries to drastically reduce their consumption patterns. Again, narrow anthropocentrism and concern for urban pollution issues prevailed; the simplistic message presented by the media seemed to be that we can solve our environmental problems by recycling.

Academic radical environmentalism thrived, however, during the period of the 1970s and 80s as ecocentric environmental ethics was refined, and as Arne Naess and others developed a philosophical basis for ecocentrism in the deep ecology movement.

Earth First!: A case history in radical environmentalism

EARTH FIRST! AROSE in the early 1980s as a reaction to the anthropocentric pragmatic interest-group politics of mainline reform environmentalism, especially over political compromises being made in classifying public land as wilderness. Dave Foreman, Bart Kohler, Howie Wolke and other founders of Earth First! had been, in some cases, staff members of the Wilderness Society and similar organizations and were themselves professionally involved in these politics. They were shocked and disillusioned by the political compromising away of America's last wild places. Most of these people were well-versed in the history and philosophy of environmentalism and the science of ecology.

As the newest branch of ecology, conservation biology, arrived on the scene in the 1980s, their worst fears were confirmed. Michael Soulé and the other conservation biologists were demonstrating that existing legally protected wilderness and other wildlife preserves around the world were too small and the boundaries were not ecologically drawn; as a result, natural evolution and continued speciation for many species on the planet had ground to a halt.[9]

Heavily influenced by Edward Abbey's ecocentric novels which dramatized a monkey-wrenching type of activism, Foreman and the others saw the need for a revival of a radical environmentalism of the Muir- Brower no-compromise type. About the same time, Arne Naess' articulations of the ecocentrism of the deep ecology movement were becoming well known, and many in Earth First! saw themselves as an activist wing of the Deep Ecology movement. As is well known, Earth First! was highly successful throughout the 80s in its ecological campaigns and in publicizing its ecocentric philosophy, and was attracting a great deal of media attention and a large following.[10]

In 1987 a major attack against Earth First! and the Deep Ecology philosophy was initiated by Murray Bookchin and his Social Ecology group.[11] Certain casual remarks by individual Earth Firsters (made, to some extent, for their shock value to drive home the out-of-balance contemporary situation of humans on the planet), such as allowing Ethiopians to starve, and AIDS as Nature's population control device, provided Bookchin with the opportunity he needed.

Bookchin was a pioneer in calling attention to urban pollution as early as 1962, but had remained aloof from traditional environmental concerns.[12] Actually, he had criticized leading ecologists (such as William Vogt, Paul Ehrlich, and David Ehrenfeld) for years, calling them reactionary Malthusians, racists, and misanthropes, for their emphasis upon the human overpopulation problem and wild species' habitat protection. In his 1987 attack, Bookchin ridiculed the idea that humanity was overpopulating the planet, devouring its resources and destroying the wildlife and the biosphere. Further, he downplayed the science of ecology, claiming that it "zoologized" humans, and proposed that the intelligence manifested in the human world of culture should take over and guide the evolution of wild Nature. As a "social ecologist" Bookchin's overriding concern has been social justice. The underlying thesis of social ecology seems to be that we cannot solve environmental problems without first solving our social problems which, in terms of practical priorities, results in an anthropocentric agenda. One prominent Bookchinite ecofeminist claimed, at a public symposium, that if a choice had to be made between Africans starving and the future of wildlife in Africa, then the wildlife would have to go. This is a far cry from Muir's dictum about siding with the bears in a war against humans.

Bookchin's attack went hand-in-hand with increasing internal dissention in Earth First! as Earth First!'s successes and high media visibility began to attract newcomers from the ranks of urban anarchists, the Rainbow Coalition, labor organizers, and others with a "social justice" background.[13] Instead of adapting to the ecocentric vision and agenda of Earth First!, they increasingly demanded in disruptive ways that Earth First! adapt itself to their social priorities. Dave Foreman tried to reassert the priorities of ecocentrism and of putting the Earth first (in terms of human population stabilization and reduction, protecting wilderness, wild species, and biotic diversity) in an *Earth First! Journal* article in 1987, but the leftist social/political agenda and tactics were soon to prevail.[14]

In the summer of 1990 the "social justice" New Guard (apparently not interested in the philosophy, science, and history of radical environmentalism) demanded that the outstanding *Earth First! Journal* minimize the publication of articles on ecology, conservation biology, and environmental philosophy, and confine itself mainly to articles describing direct actions. These developments can be reviewed in *Earth First! Journal* since 1987, but particularly in the issues beginning in May, 1990. One leader of the New Guard is reported to have said publicly that the importance of wilderness protection should be deemphasized. Another New Guard leader, in a recent Bay Area newspaper interview, accused Dave Foreman of being elitist and macho in wanting to save wilderness primarily for his personal backpacking trips.[15] It appears that the anthropocentrism of urban environmentalism now dominates *Earth First!* while the subversive message of ecocentrism, ecological consciousness, and conservation biology is rapidly being eclipsed.

The "changing of the Guard" has been accomplished: the editors of *Earth First! Journal* resigned effective January, 1991, and the *Journal* (sans "think pieces") sails along under the editorship of the New Guard, chronicling their daring-do activities in an often unpalatable style. The so-called Old Guard founders of *Earth First!* (Foreman, Wolke, and their followers), together with John Davis and the other editors, have left to start over again with a new journal (*Wild Earth*) which focuses on the original ecocentric concerns of Earth First!: wilderness, wildlife, habitat, and biodiversity, human overpopulation, and ecosophy.

The future of radical environmentalism

As for the future of radical environmentalism, I find the situation with *Earth First!* very distressing. The Green political movement throughout the world has struggled with the problem of whether it was going to embrace the Age of Ecology (the new ecocentric consciousness and paradigm) and rise above old-line political/social ideologies (neither left nor right, but out in front), or whether it would fall back primarily into the mode of leftist rhetoric and social class struggle. Many of the problems with the Green movement stem from their inability to deal successfully with this issue.

Two political scientists writing in a recent professional journal have pointed out that the "cutting edge of environmentalism" is to be found in those countries where wilderness has provided the main impetus to national Green movements, and where radical political traditions have not been exclusively industrial-Marxist and thus anthropocentric.[16]

Radical environmentalism, in my judgment, is at a major crossroads. Will the radical environmentalism of the future move beyond pre-ecological political ideologies and genuinely put Earth first in terms of the health and integrity of Gaia's ecosystems and wild species, or will the momentum of ecocentric ecological consciousness be lost as we regress back to anthropocentric ideologies and social/political agendas which fail to integrate the social problems of humans into the wider framework of restoring the health and integrity of the Earth? And will newcomers to the radical and deep long-range environmental movements be willing to educate themselves in the history and leading ideas of ecology and ecosophy, and thus open their understanding and experience to the attainment of a genuine ccocentrism and ecological consciousness, or will they carry their baggage with them and demand that existing radical environmental organizations conform to their pre-ecological ideologies, priorities, and tactics? Hopefully, aspiring radical environmentalists will engage in some serious soul-searching along these lines in the immediate future.

Notes

1. An earlier version of this paper was presented at the "Conference on Radical Environmentalism," University of California at Santa Barbara, March 1–3, 1991.

2. For histories of the development of academic radical environmentalism, see Stephen Fox, *John Muir and His Legacy: The American Conservation Movement* (Boston: Little, Brown, and Co., 1981); Max Oelschlaeger, *The Idea of Wilderness: From Prehistory to the Age of Ecology* (New Haven: Yale University Press, 1991); Roderick Nash, *The Rights of Nature: A History of Environmental Ethics* (Madison: University of Wisconsin Press, 1989); George Sessions, "The Deep Ecology Movement: A Review," *Environmental Review* 11, 2 (1987): 105–25; Warwick Fox, *Toward a Transpersonal Ecology:*

Developing New Foundations for Environmentalism (Boston: Shambala Press, 1990).

3. Oelschlaeger, *The Idea of Wilderness*.

4. Arne Naess, "Ecosophy, Population, and Free Nature," *The Trumpeter* 5, 3 (1988): 113–19.

5. For recent discussions of Muir's ecocentric philosophy, see Fox, *John Muir and His Legacy*, 43–53, 59, 79–81, 289–91, 350–55, 361; and especially, Michael Cohen, *The Pathless Way: John Muir and American Wilderness* (Madison: University of Wisconsin Press, 1984), chs. 1, 6–7.

6. For a history of the development of environmentalism and concern for human overpopulation in the 1940s and 50s, see Raymond Dasmann, *The Last Horizon* (New York: Macmillan, 1963); Fox, *John Muir and His Legacy*, 292–315.

7. Fox, *Ibid*.

8. For a penetrating discussion of the changing perceptions of the functions of designated wilderness, see Thomas H. Birch, "The Incarceration of Wildness: Wilderness Areas as Prisons," *Environmental Ethics* 12 (1990): 3-26; see also Gary Snyder, *The Practice of the Wild* (San Francisco: North Point Press, 1990).

9. For discussions of conservation biology, see O.H. Frankel and Michael Soulé, *Conservation and Evolution* (Cambridge: Cambridge University Press, 1981); Michael Soulé ed., *Conservation Biology: The Science of Scarcity and Diversity* (Sunderland, MA: Sinauer Press, 1986); George Sessions, "Ecocentrism, Wilderness, and Global Ecosystem Protection," in Max Oelschlaeger ed., *The Wilderness Condition: Essays on Environment and Civilization* (San Francisco: Sierra Club Books, 1992).

10. For discussions of activist radical environmentalism and the origins of Earth First!, see Christopher Manes, *Green Rage: Radical Environmentalism and the Unmaking of Civilization* (Boston: Little, Brown, and Co., 1990); Dave Foreman, "Earth First!" *The Progressive* (October 1981): 39–42; Dave Foreman, *Confessions of an Eco-Warrior* (New York: Harmony Books, 1991); Rik Scarce, "*Ecowarriors: Understanding the Radical Environmental Movement* (Chicago: Noble Press, 1990); and Nash, *The Rights of Nature*, 161–213.

11. Murray Bookchin, "Social Ecology versus 'Deep Ecology': A Challenge for the Ecology Movement," *Green Perspectives: Newsletter of the Green Program* (Summer 1987).

12. See Fox, *John Muir and His Legacy*, 292; for early criticism of Bookchin as a technological utopian, see David Ehrenfeld, *The Arrogance of Humanism* (Oxford: Oxford University Press, 1978), 54, 127; for further criticism of Bookchin and a defense of deep ecology, see Kirkpatrick Sale, "Deep Ecology and its Critics," *The Nation* 22 (May 14, 1988): 670-75; and Robin Eckersley, "Divining Evolution: The Ecological Ethics of Murray Bookchin," *Environmental Ethics* 11 (1989): 99–116.

13. See Bill Devall, "Maybe the Movement is Leaving Me," *Earth First! Journal* 10, 8 (1990): 6; Devall, "An Open Letter to Earth First!ers," *Earth*

First! Journal 11, 2 (1990): 30.

14. Dave Foreman, "Whither Earth First!?" *Earth First! Journal* 8, 1 (1987): 21–22.

15. For Darryl Cherney's remarks about Earth First! decreasing its concern for wilderness protection, see Howie Wolke, "Focus on Wilderness," *Earth First! Journal* 7, 8 (1990): 7; for Judi Bari's remarks about Foreman and wilderness, see Christine Keyser, "Judi's World," *Express* 13, 17 (1991): 1, 16–17, 20–25.

16. See Peter Hay and Marcus Howard, "Comparative Green Politics: Beyond the European Context?," *Political Studies* 36 (1988): 433–48.

*
**

The Perils of Conservationist Politics:
Life in the Trenches

Pete A. Y. Gunter

I HAVE BEEN HAUNTED, while writing this paper, by passages in Annie Dillard's *Pilgrim at Tinker Creek*. We may remember Dillard's very apt remarks about ordinary human perception. We see the world impressionistically, she admonishes, noting the green fringe of trees, the blue sky, a swatch of grass, a few human figures in the foreground or background. We feel at home in a world which we have constituted for ourselves out of a mixture of impressionistic gloss and sheer familiarity:

> We don't know what's going on here. If these tremendous events are random combinations of matter run amok, the yield of millions of monkeys at millions of typewriters, then what is it in us, hammered out of these same typewriters, that they ignite? We don't know. Our life is a faint tracing on the surface of mystery, like the idle, curved tunnels of leaf miners on the face of a leaf. We must somehow take a wider view, look at the whole landscape, really see it, and describe what's going on here. Then at least we can wail the right question into the swaddling band of darkness, or, if it comes to that, choir the proper praise.[1]

That is how environmentalism began for me. I didn't know what is going on here. I discovered the hard way: through experience.

How wonderfully naive I was, with an undifferentiated, fulsome love of nature and a political optimism straight out of Perry Mason and the Boy Scout Creed. I not only believed that Justice Would Triumph. I believed that ordinary citizens, by luck, pluck, and native deviousness, could band together and defeat Goliath: lumber Goliath, big wealth Goliath, political Goliath.

I believed all that. But this was in 1960, when Hubert Humphrey lived. Then there was *Hope*. Not very intelligent hope, but hope.

I remember, as a graduate student at Yale University, finding a volume in the bookstore titled *I'll Take Texas*. As I recall, the title was in puce and magenta. I hid it in a plain brown paper wrapper, sneaked it to my room where, late at night, behind locked doors, I read it. A chapter on the coming demise of the once sprawling Big Thicket stuck in my craw. (Ivy League people are not supposed to have craws, but after all, I was only a graduate student, not a real Yalie.) I managed to build up my courage and write a letter to newly elected Senator Ralph W. Yarborough (D.-Tex.) urging that at least part of the Big Thicket be set aside for future generations. Save the Everglades of Texas, I wrote. Save the Big Thicket!

The big what? The Big Thicket, a wilderness/semiwilderness in Southeast Texas. On a map you would see it northwest of Beaumont and northeast of Houston: from thirty to eighty miles from the Gulf Coast. As a boy I fished and hunted there with a friend who had relatives in the area.

True sons of Texas, we were a menace then to anything that moved and most things that didn't. I possessed only a single shot Winchester .22 rifle. But with that rifle and ten boxes of hollow point, long rifle shells, I could, and did, decimate whole populations of hawks, woodpeckers, snakes, chickens, armadillos, and even an occasional squirrel. In those days we knew what it was like to be a real Man.

As it turns out, the Thicket is a kind of ecological ark, possessing more biological diversity per square mile than any area its size in the world—or so the Cornell University biology department informs us. Dry sandlands and wet tupelo swamps, coastal prairies and river floodplain forests, rolling piney hills and palmetto flats, baygalls, seeps, fern valleys, live there cheek by jowl in great profusion, supporting tropical, temperate, eastern and western vegetation, birds, insect life: A thousand varieties of flowering plants. Over thirty species of ferns and as many of orchids. Four species of insectivores. Once covering over three million plus acres, timbering, towns, roads, dams, oil fields and oil pipelines had reduced the Thicket to around 400,000 acres by the time I happened onto it with my lethal .22.

But to reiterate:

I knew none of this. At most I realized its folklore: Old stories of escapees from the state penitentiary at Huntsville running for the Thicket ahead of the baying hounds, of people still living in back places in log cabins with coal oil lanterns, of Confederate deserters hiding out by the sloughs and backwaters safe from conscription officers—and from Shiloh, Chicamauga, Vicksburg. Of bear hunts. Of treed raccoons on moonbright October nights.

Yes, it was hard to miss the folklore.

Only, I didn't know to call it that.

But to return to activism. To my amazement Senator Yarborough replied: not only replied but announced that because of a newly passed law he was able to set aside wilderness in Texas' national forests. He informed me that approximately 2000 acres of Big Thicket vegetation was being set aside by his request in the Sam Houston National Forest,[2] and urged me to go immediately to Texas and meet several high national forest officials who would be writing to me. They would take me on a special tour of the Big Thicket.

You could have knocked me over with a pine needle. What could I say to those foresters? How could I distinguish the Big Thicket Vegetation the chief forester referred to in his letter from any other kind of vegetation? Forests looked much alike to me; so, beyond a minimal commonsense knowledge, did trees. I'd know a bear if I saw one. But the forester didn't mention bears. Besides, they were now extinct in East Texas.

So I ducked the Senator's suggestion and the Chief Forester's invitation, and went out to get an education instead. One of those people mentioned in Mary Lasswells' book was Lance Rosier, a home-grown conservationist from Saratoga, Texas who had spent years leading writers, scientists, and garden clubs into the Big Thicket. On the way back from Connecticut to Texas I stopped in at Saratoga and befriended Lance, and followed him around Pine Island and Bayou and Village Creek and Wildcat Slough, writing down in a small spiral notebook whatever he could tell me—plenty, more than I could grasp— about the Thicket, past and present. About orchids, and resurrection ferns, and the lore of backwoodsmen, and hawks, and golden spiders, and a thousand other things. A couple of dozen lessons with Lance, on trips between Yale and then Alabama, and then Tennessee to Houston and back left me stuffed with data, fogged with folklore, and determined to do what I could. For Lance had seen the wilderness receding, fading, needlessly dying. And he said so.

I began as a writer: first with reviews and short articles in *The Living Wilderness*, *Florida Naturalist*, *Kountze News*, *Texas Observer*—whoever would publish. In due time I joined the newly founded Big Thicket Association, began attending meetings, became involved in projects, and finally ended up— to my considerable amazement—leading a statewide and finally national crusade to save the Big Thicket.

This speaker will pause for an apology. He has already said too much about himself. Pete Gunter isn't the issue. The environment is. But hopefully by setting the scene one can give a certain reality to the Noble Sentiments and Pollonian Advice that are about to be emitted.

A real environmental campaign is exhausting and confusing. If there ever were an example of William James' "blooming budding confusion" this is it. With much help from many friends I was to give over two hundred speeches and more interviews than I could count, and to make dozens of media appearances. The frenetic activity stretched from the "Today Show" to the Pilot Point Kiwanis Club, from the Houston Philosophical Society to the Corsicana Junior High School Assembly Program. Petitions, boycotts, hearings, dinners with politicos, letters, trail rides, barbecues, arguments with lumber pr men, helicopter rides with cameramen.

Well. You get the picture.

The end result, if there is an end, was the creation, in 1974, of an 84,550 acre Big Thicket National Preserve: the first such entity in the history of the National Park Service. That was of some importance. Nineteenth-century parks were created primarily for scenery. This was the first area to be set aside purely and simply for biology/ecology. For every mold. For every lichen. For every insect and insectivore.

Now, please, I ask your indulgence. It is Moralizing Time. What follows will be ponderous and very heavy. Take a deep breath. Stretch. The exits have been locked.

First come the Environmental Transcendental Postulates. Then a Decalogue: Ten Ecological Commandments guaranteed to edify and instruct.

Then . . . but that's enough.

Postulate #1: Don't be naive. You *are* being followed. In my case, and that of several others, lumber company pr men followed up every speech we gave with subsequent speeches designed to put out any fires we may have started. My publisher was pressured not to publish a book I authored, *The Big Thicket: A Challenge for Conservation*.[3] In later newspaper articles he stated that he was threatened with bodily harm. As we know, death is Nature's way of saying: Slow Down.

(John Jenkins published anyhow.)

Postulate #2: They are rich and you are poor. They have glossy pamphlets; you have a mimeograph machine. They have a whole stable of public relations personnel supported full time with lush expense accounts; you rely on "volunteer" workers (i.e., mainly, you do everything yourself.) You wear out your car. They buy five new ones. And they won't give even one of them to you.

Postulate #3: They are strong and you have only Nuisance Power. They can make sizeable political donations; you can write letters to congressmen.

Corollary: Just because you are weak does not mean you are powerless. Student boycotts at the University of Texas at Austin led the board of one lumber company (based in New York) to seriously consider selling their Texas holdings in the Big Thicket.

The following ten imperatives are universal, apodictic and darned near synthetic a priori. They are not deduced from the postulates. Hardly anything is.

I know about these imperatives because I have broken every one of them. They might be considered the result of rich practical experience and mature reflection. In fact, they are the bitter and inalienable fruits of Sin.

1. You can never know enough. You can never know too much.
   a. About nature.
   b. About political, economic, and social landscapes too.
2. Listen. Not all you will hear is true. But you'll find out which way the wind blows.

3. Always compromise. But *never compromise too soon.* If you are lucky there will be some "extremists" proposing solutions far beyond what is possible. With their help you can pose as "moderate." And achieve more.

4. In almost every case you will be dealing with politicians, from county judges to U.S. senators. Get to know them. But,

    a. Don't shout at them. They want clear arguments, not decibels.

    b. Entertain them. A cool bourbon-and-branchwater speaks volumes.

    c. *Don't let them off the hook.* Thank them *after* they have passed a bill, not during or before. If they've already got big applause, they tend to slack off.

5. Study the media. Get to know environmental reporters. Learn to use "speak out" columns. Choreograph letters to the editors. Get on talk shows. Don't be afraid to try for national publicity.

6. Get a rationale. *Why* do you want to save something? Or change something? Vague aspirations, childhood familiarities, unsubstantiated or barely substantiated claims won't cut the mustard. Polish that rationale. Add to it where helpful. Keep stating it.

7. Speak with one voice. If your cause seems to speak with contradictory voices your opponents will make the most of it. (In the case of the Thicket an umbrella group called the Big Thicket Coordinating Committee was formed to create consistency. Its twenty-three organizations ranged from the Houston Sportsman's Association to the Sierra Club.)

8. If you make a compact with a politician—when you finally come down to brass tacks—*don't back down on it.* Political people take a dim view of this. And they never forget. An agreement may seem casual enough at the time; once made, however, consider it set in concrete.

9. If you are dealing with an issue that involves a particular locale: *Always* find local people who are on your side. Spare no effort in this. In the eyes of Congress nothing counts for more than local support: even in another congressman's district. (Recently we were able to take a dam upstream from the Preserve entirely off the U.S. Corps of Engineers' list of potential reservoirs— partly by organizing local people.)

10. The game ain't over when the fat lady sings. Once you've gotten what you want through government—if that is how you get it—then you have finished one task and begun another. In the case of the Thicket, as in most of its kind, after passage, boundaries had to be fixed, appropriations gotten, management plans worked out, a visitor's center placed and built, public roads rerouted—and this hardly scratches the surface. In short: You were an advocate; now you are a watchdog. The categorical imperative of all categorical imperatives is: *Don't weaken.*

I would like to conclude with two brief remarks. The first concerns the game that fails to end when the lady sings. Today we are still struggling to complete the Big Thicket National Preserve. A bill is pending to add a forty mile stream corridor (Big Sandy-Village Creek Corridor) to the Preserve, connecting several otherwise isolated Preserve units. Thirty-one years after a stammering, poorly typed letter to Ralph Yarborough, the effort continues.

The second concerns the limits of a talk like this. I have been asked to write about practical politics, and to do so is inevitably to leave much out: to draw too narrow a focus. More broadly, I now believe that environmentalists, if they are to reach the people, must find a new, more populist, less elitist rhetoric. More generally still, I would like to insist that we be not only politically astute environmentalists but deep ecologists as well. We need never to get lost in the details of tactics and the heat of battle. We need to remember the broader, indeed, the broadest view. There is a sacredness about the Earth. We live, and we struggle, to honor it.

NOTES

1. Annie Dillard, *Pilgrim at Tinker Creek* (New York: Harper and Row, 1988), 8–9.

2. This area later became part of the Little Lake Unit in the Sam Houston National Forest: a piece of wilderness set aside from the clearcutting, bulldozing, burning and brush-hogging of the National Forest Service. Cf. Edward C. Fritz, *Realms of Beauty: The Wilderness Areas of East Texas* (Austin: University of Texas Press, 1987), 116.

3. Pete A.Y. Gunter, *The Big Thicket: A Challenge for Conservation* (Austin: Jenkins, 1973), 174.

＊
＊＊

PART TWO

ENVIRONMENTAL SCIENCE
TODAY AND TOMORROW

*
* *

Conservation Biology
and Sustainable Societies:
A Historical Perspective

Curt Meine

ON THE FIRST EARTH DAY, so the story goes, a group of concerned citizens in the San Francisco Bay area decided to make a statement about the pervasive environmental effects of the automobile in our society. They buried one. In so doing, they symbolically interred all the pollution-burdened skies, soiled seacoasts, lost wildernesses, bulldozed neighborhoods, gridlocked cities, *ad infinitum* urban sprawl, ostentatious wealth, unscrupulous mechanics, and drive-in churches that the automobile represented to them. Earth Day 1970 was nothing if not fresh.

Afterwards, it is said, on the other side of town, the leaders of a money-strapped community assistance organization vehemently objected to the display. They protested, quite naturally, that they could have used the car to transport the elderly and deliver food to the needy.

It was so much easier to be environmentally aware in 1970. One sensed danger keenly, as would a wild creature. The causes were clear, the solutions self-evident. One felt the righteousness of the crusade in one's bones, and knew that the simple answer was simplicity itself. In nature, one could find order, peace of mind, and the foundations for a philosophy that would save one's soul. Greed was the root of the evil, and all corruptions would wither once an awareness of ecology and a change in values pervaded society. The problems of human society? Return to nature, and all would be well.

And it was so much easier *not* to be environmentally aware in 1970. The threats to decent and durable human societies were strictly political and economic in nature, and once the impediments to political and economic freedom were removed, all would be well. Insufficient and inequitably apportioned wealth was the root of evil, and all corruptions would diminish once the economic pie was made larger and/or (depending upon one's political philosophy) more fairly distributed. The environmentalists were not simply unrealistic; they were irrelevant at best, dangerous at worst—a strange collection of irrational utopians, discontented Luddites, social misfits, misguided misanthropes, and probably atheistic anarchists, supplemented by the odd scientist or two. Ecology? What was that? Nature? No need to deal with bothersome complications. The *real* problems of the *real* world would be solved through economic expansion and restructuring.

Ah, for the days of stereotype and dogma, simple answers and narrow definitions. They demanded so little of us. They lent themselves so much more easily to rough rhetoric, bloody politics, and the massed media. They were so much less of a drain on our psychic energies than lifelong devotion to what Edward Abbey called "Reason with a capital *R*—Sweet Reason, the newest and rarest thing in human life, the most delicate child of human history, . . . intelligence informed by sympathy, knowledge in the arms of love."[1]

The polarity that marked American environmental attitudes in 1970 has not disappeared, but the sheer weight of environmental pressures, if nothing else, has deepened the discussion. In certain circumstances—the efforts to protect North America's remaining old growth forests and the fate of the Arctic National Wildlife Refuge in Alaska, to cite two obvious examples—the polarity remains potent, and for good reason. On these far borders, we act out the final scenes of a very old drama. Here our culture's unresolved tensions, conflicting values, and divided way of life play out on the largest remaining intact corners of the original American stage.

More broadly, the conflicting points of view can now be heard on a global scale. Developed nations grow increasingly aware of the full costs—environmental and socioeconomic—of the trail they have blazed over the last few tumultuous centuries. Developing nations naturally protest that environmental concern is a luxury of the already prosperous, and that more immediate concerns preclude such investments of time, money, and human ability. The wealthy nations, having driven the sleek automobile, are beginning to warn of its dangers. The poor nations seek the life that the car has come to represent.

But Sweet Reason, rare, demanding, and delicate though it is, is also a caustic agent. Eventually it erodes even the most granitic stereotypes and stone-faced dogmas. An appreciation of history, complexity, diversity, and humanity must inevitably seep into the cracks in the solid edifice of ideas. It penetrates hard attitudes in due time—though, it seems, only after those attitudes have resulted in eroded ecosystems, degraded land-scapes, and desperate societies.

Of course, the stereotypes and dogmas were never so simple to begin with; there were just too few souls intrepid enough to explore the underlying premises and conditions. One of the few, George Perkins Marsh, warned in 1864 that "the earth is fast becoming an unfit home for its noblest inhabitant, and another era of equal human crime and human improvidence . . . of like duration . . . would reduce it to such a condition of impoverished productiveness, of shattered surface, of climatic excess, as to threaten the depravation, barbarism, and perhaps even extinction of the species."[2] The inhumane assessment of a misanthrope? The wail of a ranting Luddite? Having devoted himself to careful study and documentation of the processes by which "man had changed millions of square miles, in the fairest and most fertile regions of the Old World, into the barrenest deserts," Marsh was among the first to concern himself with what we now awkwardly call "sustainability." More poetically, Marsh challenged his contemporaries "to renovate a nature drained by [human] im-providence of [the] fountains which a wise economy would have made plenteous and perennial sources of beauty, health, and wealth."[3]

Few in Marsh's time accepted the challenge. Another "era of human improvidence" did indeed ensue, one ultimate result of which, a century later, was an Earth Day spectacle of buried automobiles and polarized attitudes. The larger questions, questions as pervasive and encircling and ignored as air itself, went largely unanswered in 1970, and we are still working on them. Put bluntly: how did we manage to work ourselves into so damn stupid a bind? Put personally: who are the bums who put us in this position? Put more soberly: has the very idea of freedom, the noblest dream of earth's "noblest inhabitant," been reduced to the mere opportunity to choose between two sides of the same corrupt coin—progressive environmental degradation and a progressively "improvident" human society?

One of those who would later accept Marsh's challenge—Aldo Leopold—once wrote that "conservation, without a keen realization of its vital conflicts, fails to rate as authentic human drama; it falls to the level of a mere Utopian dream."[4] Leopold was writing specifically in reference to conservation education, and its need to confront dilemmas directly and realistically, not merely lapse into depressing statistics and rehashed propaganda. But more generally, Leopold's point was directed toward all who strove so mightily toward simplistic personal utopias, but who neglected to respect or replenish the "fountains" to which Marsh alluded: the firmament, the cyclic waters, the common ground, the protective forests, the wetlands, grasslands, rangelands, and deserts, the incalculable floral and faunal heritage—the material world of which all dreams are really made, and on which all dreams will always depend.

***

LEOPOLD POSSESSED as keen a realization of conservation's "vital conflicts" as any person of his day. His whole life may be read as an effort, not to achieve ultimate resolution of those conflicts—he was too pragmatic a person to believe that final resolution was possible—but to communicate the urgency of conservation, to broaden its meaning, to contribute to its scientific foundations, to extend and strengthen its rationale, and, not least, to accentuate the positive impact of conservation not only on the natural world, but on human individuals and human communities. By such means was Leopold himself able to come to grips with conservation's conflicts, and help us to understand the wide gap in human consciousness that precipitates them.

This concern with ultimate sources of conservation disputes reverberated in Leopold's writing throughout his career. He gave especially poignant expression to it in one obscure, fragmentary manuscript: a three-paragraph introduction to an essay, scrawled on the back of a piece of hotel stationary. The hotel was in Berlin, where Leopold was staying during an extended tour of Germany and neighboring lands. The year was 1935, and Leopold had undertaken the tour to examine the history and status of forestry and wildlife conservation in central Europe. I have always imagined the scene to be cold, confining, uneasy: Aldo Leopold, his conservation philosophy forged in the free-wheeling and spacious American context, sitting quietly at a desk in a dim hotel room in Nazi-era Berlin, fully cognizant of the political forces then upheaving Germany, trying to come to terms with the seeming irrelevance of conservation under such circumstances. He reached back in time in an attempt to expand the context:

> The two great cultural advances of the past century were the Darwinian theory [of evolution] and the development of geology. The one explained how, and the other where, we live. Compared with such ideas, the whole gamut of mechanical and chemical invention pales into a mere matter of current ways and means.

Then he focused on the present and his own most absorbing interest:

> Just as important as the origin of plants, animals, and soil is the question of how they operate as a community. Darwin lacked time to unravel any more than the beginnings of an answer. That task has fallen to the new science of ecology, which is daily uncovering a web of interdependencies so intricate as to amaze—were he here—even Darwin himself, who, of all men, [would] have the least cause to tremble before the veil.

And then Leopold cast his thoughts toward the future:

> One of the anomalies of modern ecology [by which Leopold probably meant, more precisely, *applied* ecology or *conservation*] is that it is the creation of two groups, each of which seems barely aware of the existence of the other. The

one studies the human community almost as if it were a separate entity, and calls its findings sociology, economics, and history. The other studies the plant and animal community, [and] comfortably relegates the hodge-podge of politics to "the liberal arts." *The inevitable fusion of these two lines of thought will, perhaps, constitute the outstanding advance of the present century.*[5] [Emphasis added]

Leopold left the fragment unrefined and the essay incomplete. The statement, as it stood, was uncharacteristic. Such sweeping predictions were not generally Leopold's style. It does, however, reflect his typical unwillingness to treat the human community and the natural community—and, analogously, the liberal arts and sciences—as "separate entities." It reflects, too, his dawning sense that this split was destined to be of short duration, not necessarily (to paraphrase one of his own later statements) because it was bad for the plant and animal community, but because it would finally be bad for people.[6]

Has Leopold's prediction come true? Have the "two lines of thought" moved toward a fusion point? The evidence on Earth Day 1970 was certainly mixed. That season, one could hear Richard Nixon declare a clean environment "the birthright of every American," while the *New Republic* disparaged the environmental movement as "the biggest assortment of ill-matched allies since the Crusades."[7] Two decades—each proclaimed "the decade of the environment" at one point or another—have since passed, and yet another "decade of the environment" has begun. As if all previous decades had nothing to do with "the environment." As if in the future, once we have succeeded or failed in our efforts, we will enter a decade *not* "of the environment."

The question can be approached from another angle: instead of looking for direct evidence of fusion, we can search for circumstantial evidence of attitudes that are less polarized, of a stronger foundation of shared assumptions, maybe even of that rare "intelligence informed by sympathy, knowledge in the arms of love." Here there is cause for hope. Far away from both the hard edges and the alleged centers of power, quiet voices speak of changes that are no less (and arguably more) revolutionary for being local, personal, and incremental.

A ridge-tilling Illinois farmer: "I wake up in the morning and I have choices. What do I want to do today? I want to grow crops and turn them into money. But I don't want to handle pesticides if I can help it. If I gotta do pesticides, then farming ain't fun."[8]

A suburban recycling coordinator: "Companies used to dismiss environmentalists and environmental groups as the radical fringe. Today they are hiring those same people."[9]

From a grassroots environmental group in Poland: "We find Poland being hard hit by numerous crises at the same time. To think that there are different crises isolated one from another is not the best way to understand them. People have no chance to meet their basic needs. They are lacking healthy and cheap food, medicines, flats, water, air, joy, freedom, and vision. . . . It is getting grey and sad around here."[10]

A Brazilian journalist: "We should wage the environmental battle for ourselves, not for anybody else. It all boils down to responsibilities. Developed and developing countries each have different responsibilities to their people and to the world. . . . Only if North and South start learning from each other's problems and accepting their respective responsibilities to humanity will the environmental issue be properly addressed."[11]

What is happening here? For those who have been working in the environmental arena over the years, such declarations are signs of the most encouraging kind—of fellow citizens whose practical needs, concerns, and desires have led them to investigate the basic premises, confront the conflicts, and question the assumptions and the authorities. They suggest that the stereotypical attitudes of Earth Day 1970 have matured into something more substantial; that ideas about what constitutes "environmental concern" have grown beyond (without outgrowing) considerations of clean air, clean water, and wildlife; that ideas about social justice and economic well-being no longer exist in an absolute environmental vacuum; that ideas about the communities we live in, and about the very concept of community itself, are broadening.

From the standpoint of both history and contemporary environmental concerns, the apparent relaxation of the old polarity raises many questions. How deep do the convictions go? Are these signs that we are truly reaching Leopold's "fusion point"?

Are we moving toward that point quickly enough? And if our attitudes are becoming less polarized, have we merely succumbed to ineffective, but feel-good, compromise? Finally, and most importantly, will the consensus be strong enough, ingrained enough, to endure harder times, and the onslaught of cynicism that hard times might bring? In our understanding of the relationship between human and environmental well-being, have we forged links between causes and effects strong enough to withstand unanticipated social and economic pressures?

These questions lead us back to the "two lines of thought" that Leopold saw coming together: the study of the human community, and the study of plant and animal communities, or ecosystems. Since Earth Day 1970 both of these areas of knowledge have evolved in response to heightened, more widespread, and more immediate environmental concerns. Participants in the 1970 Earth Day observance would recognize most of the principles that underlie the new emphases, but might not understand today's language. As we have redefined the issues, and as the issues have redefined us, new terms have emerged to frame our discussion. This is nothing new. Throughout the history of science, conservation, and the environmental movement, new ideas have required new words, and inflexible words have been shed like snakeskins. The more usable and adaptable remain, and some are among those most familiar to us: "wildlife," "conservation," "ecology," and even "environment" itself.

Two of the more recent coinages—or at least new uses for older words—are worth examining. The scientific basis of conservation, the "study of the plant and animal community" of which Leopold wrote, is being reformulated in part by a movement toward "conservation biology." Whereas in 1970 the established disciplines were assumed to provide the sufficient scientific basis for conservation, the emergence of conservation biology in the 1980s signified the belief of many that it was necessary to refocus scientific energies, particularly with regard to a most basic attribute of natural systems: "biological diversity" or, for short, "biodiversity." And whereas in 1970 the basic durability of human communities, in all their permutations and diversity, was simply assumed, many speak now of the need for "sustainable" societies. "Sustainability" has muscled its way into the highest government circles as both a policy objective and an environmental goal. Its use implies that many believe it necessary to redefine what the human community is about.

The speed with which these terms have been adopted—they have gained in use, definition, and circulation only in the last decade or so—is indicative of the flux in our individual and collective thinking. They incorporate senses, ideas, nuances, and shades of meaning that have been present much longer, but they are by no means firmly defined or accepted. Something in recent times, however, has allowed them to flourish. Words, like plant and animal species, inhabit particular niches and evolve with time. Their linguistic ranges shrink and grow, their meanings change, their forms are shaped and tempered by circumstance and opportunity. The question for us is: what do these words tell us about our cultural environment and the shifting relationship between the "two lines of thought" in the epoch following Earth Day 1970? Do they offer insights into the changing physics of our planetary dilemma—the degree of fusion or polarity on matters environmental?

***

IN 1986 CONCERNED conservation professionals banded together to form a new organization, the Society for Conservation Biology. The founding of the new organization was one expression of growing concern over the accelerated loss of genetic, species, and ecosystem diversity around the world. "The society," noted its first president, Dr. Michael Soulé, "is a response by professionals, mainly biological and social scientists, managers and administrators, to the biological diversity crisis that will reach a crescendo in the first half of the twenty-first century. We assume implicitly that we are in time, and that by joining together with each other and with other well-intentioned persons and groups, the worst biological disaster in 65 million years can be averted."[12]

These concerns, of course, predated, but were restimulated by, the rise in popular understanding of ecology and human environmental impacts that Earth Day 1970 represented. There was, in 1970, no one interdisciplinary organization of biologists devoted to the broad range of conservation questions. The biological foundations of conservation had been built over the previous seventy years or so in a wide variety of relevant disciplines: geography, forestry, wildlife ecology, ornithology, zoology, botany, entomology, genetics, soil science, agronomy, limnology, marine biology, and so on through the catalogue of specialties. The application of biological research to conservation work, in short, already had a long and rich history. Each of these had made important contributions to the general cause of conservation, and each in turn had benefitted from advances made in other fields.

None of these fields alone, however, had provided the information, techniques, or perspective sufficient to counter the quickening trend of environmental degradation and attendant biological impoverishment. In the 1970s the expanded environmental movement attempted to confront the trend through a series of crucial conferences and legislative initiatives. The United Nations Conference on the Human Environment in Stockholm in 1972 brought the full range of environmental issues, for the first time, before the international community. The 1972 conference built on an earlier 1968 Conference on the Use and Conservation of the Biosphere, sponsored by the United Nations Educational, Scientific, and Cultural Organization (UNESCO). One result of this earlier conference was a coordinated Man and the Biosphere (MAB) Program that included the establishment of an international system of biosphere reserves, an important step in defining the issues that conservation biology would soon focus on. Another important milestone came in 1973, with the institution of the Convention on International Trade in Endangered Species of Wild Fauna and Flora (CITES), which provided scientific and administrative procedures for the protection of endangered species worldwide.

Meanwhile, in the United States, the National Environmental Policy Act (1969), and in particular its provision requiring federal agencies to prepare environmental impact assessments prior to undertaking proposed actions, would with time give professional biologists an enhanced (though rarely a central) role as environmental analysts and advisors. The necessity of interdisciplinary research on environmental impacts and policy under NEPA hastened what was already a growing realization that disciplinary expertise had to be combined with interdisciplinary coordination if conservation efforts were to succeed. In 1973 the Endangered Species Act formalized the growing public concern over the loss of native plants and animals, providing unprecedented legal mechanisms for the protection and restoration of threatened and endangered flora and fauna on federal lands. Importantly, the act also directed that the habitats of these organisms be protected, although in practice this protection has been weak. Nevertheless, these and other statutes passed in the same period gave conservationists hope that endangered plants and animals and threatened habitats would gain full consideration in the face of human economic pressures.

Translating this hope into effective management procedures would not be so simple. Through the 1970s and into the early 1980s, the old truth, confirmed throughout conservation history, had to be proven once again: that, absolutely necessary as legal measures were, they alone could not achieve conservation. Basic biology supersedes society's laws. Ultimately, attaining conservation goals depended on understanding and changing entrenched patterns of resource use and abuse that threatened plant and animal populations and habitats and disrupted the functioning of ecosystems.

Something seemed to be lacking, too, in the ability of the various conservation-related sciences, acting in isolation, to respond to these challenges, and especially to the speed and scale at which they were occurring. Whole systems seemed to be increasingly at risk. In the temperate zones, intensified land use accelerated long-term trends in the fragmentation, isolation, degradation, disruption, and outright destruction of forests, farmland, woodlots, range, grasslands, and wetlands. Aquatic systems and the fisheries they supported were increasingly taxed by declining water quality, heavy harvesting, species introductions, and short-sighted fishery management strategies. Desertification threatened already sensitive arid lands around the world, but especially in sub-Saharan Africa. And in the humid tropics, the conflagration of massive deforestation was beginning to build, drawing greater scientific attention to the prime attribute of the rainforests: the sheer diversity of the life forms they contained.

Prior to the 1980s, biological diversity was a relatively neglected concept in conservation, not so much because its importance was doubted, but because it was taken for granted. Theoretical ecologists had long debated the question of diversity and its relation to ecosystem stability—coming to no definitive conclusion other than that the relationship was intricate—while field conservationists were more concerned with practical efforts to save, protect, and manage particular species, wild places, and populations. Biological diversity, in short, was the medium in which conservation took place—so pervasive, so definitive, and so self-evident that even conservationists seemed unable to think of it with the objectivity it demanded. It was assumed. It was as obvious as life itself.

And extinction, diversity's partner in the evolutionary process, was likewise a relatively neglected area of study from a conservation standpoint. Paleontologists, of course, studied ancient extinctions, and in doing so provided the foundation of knowledge on which an appreciation of contemporary diversity could be built. The basic biology, however, of recent anthropogenic extinctions generally went unscrutinized. Perhaps this could be attributed to the difficulty that conservationists had in examining objectively a process they were sworn to forestall. "For one species to mourn another," Leopold wrote with the passenger pigeon in mind, "is a new thing under the sun."[13] Those in mourning do not immediately turn to the clinical questions of cause and effect. The possibility, however, of frequent, imminent, and prominent extinctions and extirpations—in America, the California condor and the grizzly, the snail darter and the Desert Hole pupfish, the Furbish lousewort and the blue whale—forced biologists to focus on the phenomenon of extinction *as a process*, and to do so *on a global scale*. This, in turn, led many back to the smaller, quieter, less gaudy members of the biotic community, the vast majority of which, it was pointed out, science had yet even to dignify with proper Linnaean nomenclature. Even the extent of that great unnamed majority was a matter of pure speculation—unknown, as Harvard biologist E. O. Wilson would note, "even to the nearest order of magnitude."[14]

Even as conservationists began to pay more attention to these less conspicuous members of the biotic community, they stepped up their efforts to preserve habitat at larger landscape, ecosystem, and regional scales. In the United States, this was evident through the 1970s both in the continuing efforts to preserve wilderness in accordance with the Wilderness Act of 1964 and in the steadily growing concern over the loss of diversity resulting from management decisions on national forests, rangelands, and parklands. But the preeminent example of ecosystem-level conservation efforts came in the tropical rainforests, where it became evident that the loss of diversity had already reached crisis dimensions. Although another decade would pass before the situation became highly publicized and politicized, the seeds were being sown throughout the 1970s by field biologists, ecologists, and taxonomists alarmed by the rapidly intensifying conversion of the species-rich rainforests. Perceptive conservationists and scientists had been expressing concern about tropical forests for years and even decades, but these voices had gone largely unheeded. By the late 1970s and early 1980s, the first concerted wave of articles, books, and reports appeared, raising the issue of rainforest destruction in such a manner that it could no longer be ignored.[15]

As the erosion of biological diversity became more visible and immediate, policy makers and members of the scientific community not traditionally focused on conservation came alive to the issue. In 1981, the United States Department of State sponsored an International Strategy Conference on Biological Diversity. In 1980 and 1981, the United Nations Food and Agriculture Organization and the U.N. Environmental Program hosted four conferences in Rome on the conservation of genetic resources of fish, other animals, forests, and crops. This concern over the loss of domestic animal and crop germplasm highlighted a growing trend. The loss of diversity at the genetic level, a relatively neglected area of research in conservation, became a critical focus during these years. In 1982, several conservation organizations, U.S. agencies, and UNESCO sponsored a conference in Washington devoted to the "Application of Genetics to the Conservation of Plants and Animals."[16] As a result of these efforts, the conservation of germplasm and analysis of the genetic basis of "viable" populations gained a more prominent place on the conservation research agenda. Many of those who now began to identify themselves as "conservation biologists," in fact, had their scientific roots in biology at the genetic rather than the population or ecosystem level.

The term "conservation biology" itself was not new. It had been used previously, although in an inchoate sense.[17] After 1978, the year that the first International Conference on Conservation Biology was held at the University of California-San Diego, it began to denote a more direct infusion of conservation efforts with recent findings in genetics, evolutionary biology, theoretical ecology, and biogeography.[18] At the same time, new insights in the earth and environmental sciences were enriching the intellectual atmosphere surrounding conservation issues. Since at least the mid-1970s, geophysicists had been drawing attention to the important role of rainforest vegetation in regional and global climatic patterns and the associated water and carbon cycles. Meanwhile, the revolutionary synthesis of plate tectonics in geology and the provocative elucidation of the Gaia hypothesis served to draw the life sciences and geophysical sciences closer together in both theory and method. A further contribution came in the early 1980s, when Luis and Walter Alvarez stimulated scientific discussion of mass extinction with their novel theory that the extinction event at the Cretaceous-Tertiary boundary was due to a devastating asteroid impact. The theory (and, importantly, related debates surrounding the "nuclear winter" scenario in 1984-1985) not only put biological diversity into a revised evolutionary and ecological context, but gave it enhanced visibility at both the professional and popular level.[19]

In 1986 a second international conference on conservation biology was held at Ann Arbor. By this time professional interest was focused enough to support establishment of the Society for Conservation Biology and its new journal, *Conservation Biology*. Growing numbers of scientists and conservation officials, and in particular young professionals trained with post-Earth Day environmental sensibilities, found in the approach of conservation biology a refreshing perspective on the problems confronting their varied fields and professions. More specifically, it represented an intensified effort to address conservation issues as they must inevitably be addressed: through the integration of conservation theory and practice, philosophy and policy, underpinned by solid interdisciplinary scientific research and application, involving all levels of biological organization.

These varied forces reached critical mass in September 1986, when the U.S. National Academy of Sciences and the Smithsonian Institution sponsored a four-day "National Forum on BioDiversity." The forum, which included thousands of participants linked by satellite to Washington, brought together not only prominent biologists, but anthropologists, economists, philosophers, policy makers, a poet, a filmmaker, a theologian. In 1988, the National Academy published the proceedings of the forum in the book *Biodiversity* (the abbreviated term "biodiversity" was suggested by Walt Rosen, a member of the NAS staff who assisted in planning the forum).[20] The volume has since become a standard reference. Its editor, E. O. Wilson, summarized the basic intent of the forum in his opening paragraph: "Biological diversity must be treated more seriously as a global resource, to be indexed, used, and above all, preserved. . . . We must hurry to acquire the knowledge on which a wise policy of conservation and development can be based for centuries to come."[21]

This, to a large degree, would become the mission of those who adopted the label "conservation biologist." The new terminology has not been without its critics, particularly in "traditional" conservation fields such as forestry and wildlife management, and this criticism has not been without some merit.[22] New terms always run the threat of becoming buzzwords—substitutes for, rather than indicators of, progress. But terms come into (and fade from) circulation for good reasons, and these reasons are usually more important than the terms themselves. In this case "conservation biology" and "biodiversity" capture new emphases in the science, its epistemological context, and its application: the crisis orientation; the neglected importance of diversity in previous conservation research; the need to move beyond conservation efforts that focused only on a few large, beautiful, edible, popular, watchable, huntable, and/or economically important species; the need to adopt a longer-term view of the evolutionary and ecological background of conservation practices; the recognition of the importance of geographical scale and biological hierarchies; the impact of unprecedented global environmental phenomena (e.g. population growth and the greenhouse effect) on floral and faunal composition; accordingly, the need to bring a more balanced international perspective to conservation, and to anticipate problems in exporting the traditional temperate zone/American model; and the need to counter the "hardening of the categories" that afflicts scientific (as well as humanistic) inquiry and exploration in the modern age.

Conservation, in short, is evolving, reforming itself, as it has always done in the past, as it must to meet the demands that will be made on it in the future. The emergence of conservation biology and the rise of biodiversity as a unifying concept indicate that the scientific foundations of conservation are expanding, overlapping, and shifting, out of necessity and through new understanding. In this, it is not so much a new science as it is a more comprehensive, better integrated response to problems that are themselves, tragically, more extensive, more immediate, and more intricate than most had realized in 1970. In doing so, it builds on and extends a tradition of applied ecological knowledge in conservation that has always been present, but that has long been stifled by narrower approaches. Universities, agencies, professional groups, and conservation organizations have begun to adapt, in varying degree, to the reorientation. The process continues. "These changes," as E. O. Wilson writes, "can be expected to reshape the international conservation movement for decades to come."[23]

✳✳✳

WILSON NOTES THAT, during these same years, awareness was growing "of the close link between the conservation of biodiversity and economic development. In the United States and other industrial countries, the two are often seen in opposition, with environmentalists and developers struggling for compromise in a zero-sum game. But in the developing nations the opposite is true. Destruction of the natural environment is usually accompanied by short-term profits and then rapid local economic decline."[24] This awareness grew quickly in the 1980s as the bright bloom of the Green Revolution began to fade in many areas, and as large-scale international development policies and projects, many of them profoundly threatening to biological and cultural diversity, failed to fulfill their promise. Meanwhile, the industrialized nations began to face the accumulated costs of environmental neglect—in their soils and waters, on their remaining wildlands, on their farms and in their suburbs, in their industrial sectors and urban cores. Even the wealthy nations had to begin to assess the need for what George Perkins Marsh called so perfectly, so long ago, "a wise economy."

In 1970, the blueprints for building such an economy, and a society around it, were only vaguely visible. Part of the problem was that conservationists had long had their hands full trying to save all the pieces that the older structure had marginalized; keeping abreast of losses precluded any systematic devotion to alternatives. But another part of the problem was that most social scientists and development experts had received training in sociology, economics, history, and political science that treated human society "almost as if it were a separate entity." The social sciences were not alone in this regard, of course, but merely reflected dominant social attitudes and the ever-greater specialization of the professions and disciplines following World War II. Even professional conservationists, by and large, still treated human society and the biotic community as "separate entities," simply concentrating on the latter.

The prospects, then, circa 1970, were not altogether encouraging. The biologists and conservationists who knew most about the ecological foundations of human society were generally unprepared, by training and often by temperament, to tackle the tough questions of socioeconomic reform and readjustment. Social scientists, economists, historians, politicians, and the media were unschooled in the intricacies of the natural and environmental sciences, and often uninterested. That these two broad human "taxa" could learn fundamental lessons from one another seemed a faint hope. That they would nonetheless *have to* learn from one another would, as we entered the 1990s, be beyond question, at least for a growing number of concerned citizens, scientists, and policy makers.

Conservation biologists have, as noted above, played an important role in changing the terms and the atmosphere of the debate. In doing so, they have had to broaden their traditional focus on biological reserves, habitat preservation, and single species management and come to see biological diversity in its full social, historical, and geographical context. Ray Dasmann summarized this realization in 1987, when he wrote that "we cannot extend our concern for wild nature unless we are also concerned for the welfare of people, because the two cannot be separated. . . . If efforts to conserve biological diversity are to succeed, nature conservation must become part of a total land use pattern. Sustainable use and management must have a role." Dasmann himself noted that "a decade or two ago these words would not have been received with much enthusiasm."[25] In those

intervening years, however, the increasingly evident relationship between social and economic pressures and environmental
decline, particularly in the developing world, had left little
room, and less time, for piecemeal approaches.

During these years, the term "sustainable" emerged to describe
the new orientation. As with any new catchphrase (including
"conservation biology"), it faced serious obstacles to acceptance:
the red pencil of editors, the blank stares of the unaware, the
coolness of entrenched academics, the caution of reluctant
bureaucrats, the ardor of its own adherents. Many authors ran
with it, not pausing to look back. Others lamented its lack of
definition, as if this were some fatal disease, and not the sign of
new ideas being conceived. Still others were understandably
insulted. Among development theorists, analysts, and experts,
"sustainability" was always a given. Just as conservationists
had taken biological diversity for granted, so had their colleagues
in the social sciences rarely stopped to ponder the ecological
preconditions of social cohesion and continuity; it, in a sense,
was the medium in which *their* work took place. Now, however,
those basic factors had to be given greater weight.

These considerations were hardly being weighed for the first
time. Thomas Robert Malthus in many ways began the discussion
in 1798 with his *Essay on the Principle of Population*, applying
to human society a primitive version of the principle that
ecologists would later call "carrying capacity." Malthus' scenario
was not optimistic. Extrapolating the trends as he saw them, he
discerned an inevitable tendency toward overpopulation and
resource scarcity. Marsh, looking in 1864 at the influence of
prevailing patterns of resource use on "the social life and social
progress of man," was a bit more sanguine; while he held only
a "faint hope that we shall yet make a full atonement for our
spendthrift waste of the bounties of nature," he did look forward
to "an epoch when our descendants shall have marched as far
beyond us in physical conquest as we have marched beyond the
trophies erected by our grandfathers."[26]

With the subsequent rise of the conservation movement in
the United States in the early 1900s, the notion of sustainability
appears in the guise of "sustained yield" of timber, water supplies,
range forage, fisheries, and, somewhat later, wild game populations.
Compared to the lack of restraint that had characterized the use

of these resources over the centuries, the idea that they could and should be managed rationally for perpetual economic benefits was revolutionary. But the underlying reductionist, utilitarian assumptions—that they were discrete "resources," and that they *existed* solely for the perpetual economic benefit of human beings—would come to be challenged in turn by a yet broader view of conservation that combined science, ethics, aesthetics, and long-term economics in its rationale.

An important turning point came in the 1930s, when a vanguard of American scientists and conservationists began to explore the social and economic as well as biological ramifications of the emerging science of ecology. The influences were many: the intensification of agriculture, particularly in the American midwest, in the 1920s; subsequent accelerated rates of soil erosion and habitat loss; the devastation of the Dust Bowl; the simultaneous disruption of the human economy and social system; the progressive disappearance of wilderness; unprecedented irruptions of wildlife populations; increasing rates of local extinctions and extirpations.[27] These and other factors combined to produce a realization that what we now term "sustainability" was in fact predicated not merely on the fine latticework of ideas concerning the human community, but also on basic ecological facts of life.

Aldo Leopold embodied this transition. A product of the early conservation movement, his lifelong efforts to reconcile the preservationist and utilitarian viewpoints within the movement, and to bring a stronger scientific viewpoint to environmental dilemmas, led him to question the very premises on which conservation had been built. Ultimately, he would argue for an approach that stressed the need for both preservation and "wise use" within a broader context, based on a critical understanding of environmental change and ecological functions, an informed aesthetic appreciation of natural processes and objects, and an ethical regard and respect for the "land community" as a whole.[28] The transition did not come quickly for Leopold, and it did not come easily. The aspect of his shift in perspective most pertinent to the current discussions of sustainability occurred in the latter half of the 1930s. In 1935, Leopold noted that "philosophers have long since claimed that society is an organism, but with few exceptions they have failed to understand that the organism includes the land which is its medium."[29] By the end of the decade that essential relationship would be recast in his thought and in his prose: ecology, the "new fusion point for the sciences," taught Leopold to see land as the whole that included human society, and the study of its lifeways could help reveal how human beings might live decently within it. He charged his scientific brethren with the task: "We might get better advice from economists and philosophers if we gave them a truer picture of the biotic mechanism."[30]

That task, however, would be put on hold by World War II. The struggle for dominance and survival in the human community drained attention away from the broader considerations of the relationship between the human and biotic communities. The promising syntheses within conservation in the late 1930s faded in the fog. As Leopold himself noted, "Against a background of war, [conservation] looks like a milk-and-water affair."[31] But in ways not immediately apparent, the war also served to focus attention on conservation issues. The internationalization of a generation, the heightened role and responsibility of science, the accelerated pace of resource exploitation, the advent of new production processes, and the employment of atomic weaponry— the most concentrated threat to "sustainability" yet devised— would redefine the context of environmental debate.

In the years immediately following the war, a series of ecologically informed, international-scale assessments of the global condition appeared that were, in Robert Paehlke's words, "precursors of a coming change in public consciousness."[32] Two of these, Fairfield Osborn's *Our Plundered Planet* and William Vogt's *Road to Survival* (both published in 1948), were especially important in conveying the message to the conservation community. Osborn, bringing a more modern perspective to Marsh's essential points, wrote that "man must recognize the necessity of cooperating with nature. He must temper his demands and use and conserve the natural living resources of this earth *in a manner that alone can provide for the continuation of civilization*. The final answer is to be found only through comprehension of the enduring processes of nature."[33] "Above all," Vogt wrote, "we must learn to know—to feel to the core of our beings—*our dependence upon the earth and the riches with which it sustains us*"[34] [emphases added]. The same intellectual climate led to

Leopold's concurrent expression of the land ethic: "A land ethic changes the role of *Homo sapiens* from conqueror of the land community to plain member and citizen of it. It implies respect for his fellow members, and also respect for the community as such."[35] Such expressions, ignored for the most part by the intellectual leaders of the era, and little heeded in the post-war rush to normalcy and prosperity, nonetheless would have a tenacious hold on readers who, by one route or another, came to them. Those readers, in turn, would lead the country to Earth Day.

The process by which the various strands of "sustainability" came to be woven together in the ten years before and after Earth Day is a complex one, and we are only now gaining perspective on the patterns. The emergence of concern about biological diversity, summarized above, was a highly significant part of that process. Others bear at least brief mention. With *Silent Spring*, Rachel Carson brought the basic concepts of ecology to an audience of unprecedented size. Paul Ehrlich's best-selling book *The Population Bomb* (1968) picked up the Malthusian theme and reignited that aspect of the discussion. Garrett Hardin's essay "The Tragedy of the Commons," first published in 1968, showed the power of carrying capacity as a conceptual tool in comprehending the environmental impact of human population pressure. Barry Commoner's work in the 1960s and 1970s emphasized the role of technological choices and socioeconomic systems in determining the actual impact that population growth would have on environmental systems. The energy crises of the 1970s brought home the fundamental role of energy production, transport, and consumption in determining not only the environmental impact, but the very character, of modern society. Amory Lovins' *Soft Energy Paths* (1977) led the way to alternative analyses of this critical factor. Kenneth Boulding, E. F. Schumacher, and Herman Daly, among other non-traditional economists, helped to move that field toward not simply a more rational approach to ecological conditions, but toward an economics that drew on and applied basic ecological principles to human economic systems.[36]

This partial list of the best known publications can only serve, at best, as an index of the general surge of environmental thought and activism. One might examine any specific area of environmental reform to trace the impact of this surge, but the one that would prove most significant in the evolution of the "sustainability" idea would be the most fundamental: agriculture. As these assorted factors in the formula—population growth, technological change, pollution control, energy use, economic analysis—jostled for accommodation in the 1970s and the connections between them became better defined, American agriculture in particular presented an important area for their integration.

Agriculture, in a post-World War II pattern that characterized, to a greater or lesser extent, all areas of resource use in the developed world—forestry, fisheries and water resource management, range management, wildlife management, recreation—became more specialized, intensive, and commodity-driven even as many aspects of the resource base deteriorated. In each of these areas, alternative, more integrated approaches went into deep eclipse for a generation. On the agricultural front, many traditional techniques of soil conservation and management were widely abandoned, but they had secured a niche in modern agricultural policy and science beginning in the 1920s and held on to that niche through the post-war decades. In addition, the strong tradition of more holistic approaches to farming, though overshadowed, never disappeared entirely. With the rise of ecology these approaches began to gain the attention and imprimatur of science.[37]

The manifold impacts of industrial-style agriculture— groundwater contamination, accelerated soil erosion, soil compaction, nutrient depletion, salinization, pesticide resistance, diminished soil floral and faunal activity, decreasing wildlife populations, loss of crop germplasm diversity, farmer health problems, declining rural communities, economic insecurity— became increasingly evident in the late 1970s and 1980s, reinforcing one another in a cycle that fragmentary approaches could not effectively address. Agricultural science, education, and policy, locked into strong traditions of their own, responded slowly to the situation.

The result was a grassroots "sustainable agriculture" movement (although it had and has many other names) that brought the term into wide usage among a constituency that—importantly— could apply it on the ground and give it greater definition. This trend was reinforced when the economic crises and droughts of the mid- to late 1980s drove many farmers off the land, and forced an even greater number to take a hard look at the economics, in particular the high cost of purchased inputs, of their farming practices. As a consequence more and more farmers have begun to adopt alternative practices for economic reasons, but have realized other benefits. Sustainable agriculture has by no stretch of the imagination won the day, and the formula for achieving it will never be finally determined (and indeed must vary by site and circumstance). But the very process of research, demonstration, and application, usually led by the farmers themselves, has given strength to the movement. Along the way the term "sustainability" and its variants have come to signify a broadened approach that simultaneously tries to deal with the social, economic, agronomic, environmental, and intergenerational costs of conventional agriculture.

Meanwhile, a parallel shift has taken place—to what degree it is hard to measure—at the international level. International development programs, which by the nature of their mission entail a large component of *agricultural* development, have for decades been dominated by efforts to export the American model of high-input production agriculture. The technologies that collectively comprised the Green Revolution succeeded in raising raw yields in many regions of the world, but not without significant social and environmental costs. As the gains in yield show signs of plateauing, additional arable lands have grown scarce, and the quality of the soil and water resource base has suffered the effects of widespread neglect and abuse. The environmental and associated cultural impacts of the large scale, energy-dependent methods have called into question many basic assumptions of the development model.[38]

At the same time, the complicity of international development agencies (in particular the large international lending institutions) in funding environmentally insensitive development projects, backing economically perverse policies, and leading developing countries into deeper and deeper ruts of debt, reinforced the sense that alternative approaches were needed. The most dramatic and well publicized consequence of the old approach—the intensified conversion of the rainforests of the Amazon basin—represented the final absurdity of a way of thinking about human communities and their development that by the late 1980s had run its course. In this environmental crucible, concerns over biodiversity loss, poverty, the decline of indigenous cultures, and inappropriate approaches to development melded together in sadly spectacular fashion.

Importantly, this took place just as additional global scale environmental concerns—the depletion of the Earth's ozone layer, the specter of the greenhouse effect, the Law of the Sea negotiations—came to the fore. A culminating point in this period was the release in 1987 of the widely publicized report of the World Commission on Environment and Development, *Our Common Future* ("the Brundtland Report," as it came to be called), which called for a redirection of international development policies to foster sustainable economic development and resource management.

The Brundtland Report, however, was only the latest, most prominent (and by no means least controversial) summary statement concerning international development and environmental policy. The advocates of sustainability had been busy for over ten years in an intensive effort to define and refine their alternative analyses. Building in particular on the insights (and faults) of the landmark Club of Rome report *The Limits to Growth* (1972), this work was reflected in an outpouring of new publications that brought the idea before policy experts and the general public. A far from complete list of these publications would include: *Alternatives to Growth: A Search for Sustainable Futures* (1977), *The Dispossessed of the Earth: Land Reform and Sustainable Development* (1979), *Assessing Tropical Forest Lands: Their Suitability for Sustainable Uses* (1981), *Building a Sustainable Society* (1981), *Sustainable Food Systems* (1983), *Agricultural Sustainability in a Changing World Order* (1984), *Sustainable Resource Development in the Third World* (1987), *Sustainable Development: Exploring the Contradictions* (1988), *Sustainable Development in Agriculture* (1988), *Sustainable Environmental Management: Principles and Practice* (1988), and *Fragile Lands of Latin America: Strategies for Sustainable Development* (1989).[39] There were many others, and the library is still expanding. Inevitably, this outpouring has been criticized as a mere bandwagon phenomenon. Others hold that the bandwagon has been too long in the coming. Still others point out that the mere invocation of "sustainability" does not portend any significant concern for the broader ecosystem except as it concerns human welfare, thus perpetuating the very root cause of non-sustainability.[40]

Whatever else the use of the term "sustainability" signifies, it represents a phenomenal shift in rhetoric and *potentially* in practice. Taken in sum, studies such as those provide a detailed blueprint for social change and development that was lacking on Earth Day 1970. Sustainability, like conservation biology, represents an attempt to implement a more integrated approach to the problems at hand. It has forced those who work on these issues to go back to basics, to ask fundamental questions about their methods and goals, and to take into account—quite literally— the ecological basis of secure human societies. And for some, at least, it has opened the way to a wider ethical perspective that allows ecosystems to be "sustained" not only for their human benefit, but for their own inscrutable purposes. As with conservation biology, this reorientation has forced universities, agencies, professional groups, and conservation organizations to adapt. There are signs that these adaptations will result in real changes, although the speed and depth of change have yet to be determined. That is where we come in.

THERE WILL ALWAYS be those who insist on framing any and all environmental issues in terms of "people versus trees," "jobs versus the environment," "progress versus stagnation." But people are becoming increasingly aware of the fallacies, simplistic assumptions, and myopic historical perspectives that underlie such sentiments, and are searching for viable, workable alternatives that do not involve choosing the short-term evil of individual hardship or the long-term evil of environmental decline. We all face this challenge, each in our own spheres of influence. If the prospects sometimes seem daunting, we might take some comfort in the reminder that the great naturalist Olaus Murie once offered: "Evolution is our employer."[41]

Over the last twenty years, conservation biology has emerged as an integrative area of scientific endeavor that tries to help us fathom the full diversity of life, understand the impact of human activities on that diversity, and devise means to maintain and restore it within functionally healthy ecosystems. To do so, it has had to pay increased attention to the human side of our dilemma: the historical patterns, social conditions, policy decisions, and development goals that so largely determine the fate of biological diversity around the world—and in our backyards. Conservation biology, in this sense, is an important indication that we may be overcoming the dichotomy that Leopold described in 1935.

Over the last twenty years, too, sustainability has become an umbrella term to guide social development in a manner that recognizes the fundamental importance of ecosystem health and the diversity of life—the ultimate source and basis of healthy human societies. Although still defined largely in human economic terms, sustainability does imply that constant attention must be given to long-term environmental quality, and shows promise that it may come to recognize the inherent value of the biotic community in which human society has coevolved and with which human society must coexist. Sustainability, in this sense, is also an indication that we have advanced toward the day when Leopold's dichotomy will be of interest only to historians.

The advent of conservation biology and the widely stated goal of sustainable societies indicate that the times have indeed changed. Both concepts attempt to encompass vast realms of knowledge and activity. We find now that these spheres, the biotic and the human, can no longer be considered separately. If on Earth Day 1970 those concerned with the health of the environment and those concerned with the health of human society generally spoke past one another, this can no longer be the case. Their words were stones, cast into a large pool of frustration. Now the concentric circles emanating from those stones have rippled widely on the water, growing more comprehensive with time, meeting and overlapping, and finally converging into a common circle of concern.

✳✳✳

Acknowledgements

I WOULD LIKE to thank Max Oelschlaeger for suggesting the topic of this paper. John Ross, Reed Noss, Kenneth Dahlberg, and Allan Mazur provided helpful comments on various draft versions.

NOTES

1. Edward Abbey, *Abbey's Road* (New York: E.P. Dutton, 1979), 127.

2. George Perkins Marsh, *Man and Nature, Or, Physical Geography as Modified by Human Action*, edited by David Lowenthal (Cambridge: Harvard University Press, 1965), 43. Originally published in 1864.

3. Ibid., 29

4. Aldo Leopold, Review of A. E. Parkins and J. R. Whitaker, *Our Natural Resources and Their Conservation*, *Bird-Lore* 39, 1 (1937), 75.

5. Quoted in Curt Meine, *Aldo Leopold: His Life and Work* (Madison: University of Wisconsin Press, 1988), 359–60.

6. See "The Outlook for Farm Wildlife" in Susan Flader and J. Baird Callicott, eds., *The River of the Mother of God and Other Essays by Aldo Leopold* (Madison: University of Wisconsin Press, 1991), 326.

7. See Roderick Nash, *The Rights of Nature: A History of Environmental Ethics* (Madison: University of Wisconsin Press, 1989), 125; and Stephen Fox, *John Muir and His Legacy: The American Conservation Movement* (Boston: Little, Brown and Company, 1981), 326.

8. "Herbicide Rates: The Quest for Less," *Prairie Farmer* 163, 4 (February 19, 1991), 14.

9. "Environment Again Seen as a Topic for Study," *Chicago Sun-Times*, November 13, 1990, 22.

10. "A Brief Version of Our Ideas," *Green Brigades–Ecologists' Paper*, 2 (Summer 1990), 15. This passage is drawn from a statement of the Congress of the Green Federation, Kosciuszki, Poland, June 30-July 2, 1989.

11. Ricardo Bayón, "The Blame? Who Cares?," in *Independent Sectors Network '92* 3 (December 1990), 2–3. Published by The Centre for Our Common Future, Geneva, Switzerland.

12. Michael Soulé, "History of the Society for Conservation Biology: How and Why We Got Here," *Conservation Biology* 1(1), 4.

13. Aldo Leopold, *A Sand County Almanac and Sketches Here and There* (New York: Oxford University Press, 1949), 110.

14. Edward O. Wilson, "The Current State of Biological Diversity," in *Biodiversity*, E. O. Wilson and F. M. Peter, eds. (Washington, D.C.: National Academy Press, 1988), 5.

15. Prominent examples from this crucial period include Norman Myers' *The Sinking Ark* (Oxford: Pergamon Press, 1979), Anne and Paul Ehrlich's *Extinction* (New York: Random House, 1981), and the National Academy of Sciences reports *Conversion of Tropical Moist Forests* and *Research Priorities in Tropical Biology* (1980).

16. I am grateful to Sherri Boykin for providing a chronology of these events in her "Brief History of the Rise of Conservation Biology," prepared for the Conservation Biology Seminar at the University of Wisconsin-Madison, February 1987.

17. Susan Jacobson notes, for example, that Paul Errington and Frederick Hamerstrom, in the first article of the first issue of the *Journal of Wildlife Management*, referred to "the new and growing field of conservation biology."

In the same issue, a statement of policy for the new Wildlife Society stated that "wildlife management along sound biological lines is also part of the greater movement for conservation of our entire native flora and fauna." See Susan Jacobson, "Graduate Education in Conservation Biology," *Conservation Biology* 4(4), 431–40.

18. Out of this initial conference emerged the volume *Conservation Biology: An Evolutionary-Ecological Perspective*, edited by M. E. Soulé and B. A. Wilcox (Sunderland, Mass.: Sinauer Associates, Inc., 1980).

19. Allen Mazur and Jinling Lee, "Sounding the Global Alarm: Environmental Issues in the National News," unpublished manuscript. I would like to thank Dr. Mazur for emphasizing these points in personal correspondence.

20. Edward O. Wilson, "Editors Foreword," in *Biodiversity*, vi.

21. Wilson, "The Current State of Biodiversity," 3.

22. See the articles, for example, in the "In My Opinion . . ." section of the *Wildlife Society Bulletin* 17(3), 335–60.

23. Wilson, "Editor's Foreword," vi.

24. Ibid., v–vi.

25. Ray Dasmann, "The Land Ethic and the World Scene," in Thomas Tanner, ed., *Aldo Leopold: The Man and His Legacy* (Ankeny, Iowa: Soil Conservation Society of America, 1987), 114.

26. Marsh, *Man and Nature*, 13, 44.

27. For overview discussions of this transition, see especially Donald Worster, *Nature's Economy: A History of Ecological Ideas* (Cambridge: Cambridge University Press, 1985; first published by Sierra Club Books in 1977), 205–290; and Nash, *The Rights of Nature*, 55–86.

28. For more detailed examinations of the development of Leopold's conservation philosophy, see Susan Flader, *Thinking Like a Mountain: Aldo Leopold and the Evolution of an Ecological Attitude Toward Deer, Wolves, and Forests* (Columbia: University of Missouri Press, 1974); J. Baird Callicott, *In Defense of the Land Ethic: Essays in Environmental Philosophy* (Albany: State University of New York Press, 1989); Roderick Nash, *The Rights of Nature: A History of Environmental Ethics* (Madison: University of Wisconsin Press, 1989); Eugene Hargrove, *Foundations of Environmental Ethics* (Englewood Cliffs, N. J.: Prentice Hall, 1989); Max Oelschlaeger, *The Idea of Wilderness: From Prehistory to the Age of Ecology* (New Haven: Yale University Press, 1991); and Curt Meine, "The Utility of Preservation and the Preservation of Utility: Leopold's Fine Line," in *The Wilderness Condition: Essays in Environment and Civilization*, in Max Oelschlaeger, ed., (San Francisco: Sierra Club Books, 1992).

29. "Land Pathology," in Flader and Callicott, *The River of the Mother of God*, 212.

30. "A Biotic View of Land," in Flader and Callicott, *The River of the Mother of God*, 267.

31. "Land Use and Democracy," in Flader and Callicott, *The River of the Mother of God*, 295.

32. Robert Paehlke, *Environmentalism and the Future of Progressive Politics* (New Haven: Yale University Press, 1989), x.

33. Fairfield Osborn, *Our Plundered Planet* (Boston: Little, Brown and Company, 1948), 201.

34. William Vogt, *Road to Survival* (New York: William Sloane Associates, 1948), 286.

35. Leopold, *A Sand County Almanac*, 204. For additional discussion of the historical and intellectual context of Leopold's land ethic, see especially S. Flader, "Aldo Leopold and the Evolution of a Land Ethic," in Tanner, *Aldo Leopold: The Man and His Legacy*; J. B. Callicott, "The Conceptual Foundations of the Land Ethic," in *In Defense of the Land Ethic*; R. Nash, "Aldo Leopold's Intellectual Heritage," in J. B. Callicott, ed., *Companion to A Sand County Almanac: Interpretive and Critical Essays* (Madison: University of Wisconsin Press, 1987); and C. Meine, "Building 'The Land Ethic,'" in Callicott, *Companion*.

36. For a full discussion of the emergence of the "sustainability" theme, see especially Samuel P. Hays, *Beauty, Health, and Permanence: Environmental Politics in the United States, 1955-1985* (Cambridge: Cambridge University Press, 1985) and Paehlke, *Environmentalism and the Future of Progressive Politics*. A helpful guide to the literature is R. W. Meridith and L. S. Z. Greenberg, "Global Sustainability: A Selected, Annotated Bibliography," *Institute for Environmental Studies Report* 137 (Madison: IES, University of Wisconsin-Madison, 1990).

37. See especially Richard Harwood, "A History of Sustainable Agriculture," in C. A. Edwards, R. Lal, P. Madden, R. H. Miller, G. House, eds., *Sustainable Agricultural Systems* (Ankeny, Iowa: Soil and Water Conservation Society, 1990), 3–19.

38. For a detailed discussion, see V. W. Ruttan, ed., *Biological and Technical Constraints on Crop and Animal Productivity: Report on a Dialogue* (St. Paul: University of Minnesota Institute for Agriculture, Forestry, and Home Economics, 1989). See also Charles Benbrook, "Protecting Iowa's Common Wealth: Challenges for the Leopold Center for Sustainable Agriculture," *Journal of Soil and Water Conservation* 46(2), 89-95. See also Richard B. Norgaard, "The Development of Tropical Forest Economics," and Stephen Gliessman and Robert Grantham, "Agroecology," both in Suzanne Head and Robert Heinzman, eds., *Lessons of the Rainforest* (San Francisco, Sierra Club Books, 1990).

39. See Meridith and Greenberg, "Global Sustainability: A Selected, Annotated Bibliography."

40. For example, see N. C. Brady, "Making Agriculture a Sustainable Industry," in C. Edwards et al., eds., *Sustainable Agricultural Systems*, 20–32; Kenneth Dahlberg, "Sustainable Agriculture—Fad or Harbinger?" *BioScience* 41(5), 337–39; Hal Salwasser, "Sustainability as a Conservation Paradigm," *Conservation Biology* 4(3), 213–16; Reed Noss, "Sustainability and Wilderness," *Conservation Biology* 5(1), 120–22.

41. Olaus Murie, "Ethics in Wildlife Management," *Journal of Wildlife Management* 18(3), 290.

*
**

Municipal Solid Waste to Fuel

Cheryl L. Brooks and
Kenneth E. Daugherty

THE PROBLEM OF municipal solid waste (MSW) disposal has become an increasingly important national issue. The average American throws away approximately 4 to 6 pounds of garbage each day. As a nation, we produce more than 165 million tons of solid waste per year, and this volume is expected to increase 20 percent by the year 2000 according to the EPA. Currently, 80 percent of the solid waste is being disposed of in landfills. However, as land space (especially in Northeastern states) decreases, disposal costs are soaring. Additionally, environmental concerns regarding soil and groundwater contamination have caused opposition to efforts to build new landfills or expand old ones. Some communities have turned to waste-to-energy incinerators as a possible solution to the solid waste disposal problem. However, mass burn incineration results in hazardous emission gases and generates toxic ash which must then be landfilled. The solution to the solid waste problem must be one which is not only effective, but also does not create any new environmental problems.

One possible answer is that of resource recovery followed by incineration. This process involves the removal at a central receiving site of recyclable materials such as glass, metals, plastics, cardboard and yard clippings from the municipal solid waste stream. These materials are then sold at their current market value. The remaining volume of waste, consisting mainly of paper, is reduced in size and compressed into small pellets known as refuse-derived fuel (RDF). RDF has an energy value of about 7500 Btu/lb, similar to that of lignite coal, and can be used as an alternate fuel in power plants, cement kilns, or other coal burning facilities without the resulting toxic ash or emissions associated with mass incineration.

In 1984 our research group began to examine ways of making RDF more chemically and biologically stable, capable of being stored for long periods of time, and more environmentally acceptable. To accomplish this goal, research was undertaken to investigate the effect of adding low-cost binders to the densifying process. Under contract to Argonne National Laboratories, our team examined more than 150 possible binders and binder combinations in the laboratory. Various tests and evaluations were conducted on RDF pellets. Information from these laboratory-scale tests and the evaluation of cost data, environmental characteristics, and binder effectiveness made possible the ranking of the binders and binder combinations. These binders included glues, oils, plastic cements, paint sludge, etc. The binders were subjected to a wide variety of environmental protocols such as toxicity properties, odor and potential emissions, as well as laboratory protocols including binder dispersability, binder Btu content, ability of the binder to wet the RDF substrate, binder ash content, pellet weatherability, pellet water sorbability, pellet caking, pellet ignition temperature, moisture content, durability, aerobic stability, and pellet fabrication costs. From these laboratory studies, the best binders seemed to be those which had large surface areas and were basic in nature. The binder of choice was calcium hydroxide or combinations thereof, which is both plentiful and inexpensive. The thirteen most promising candidates were selected for field testing at the Navy Civil Engineering Laboratory's (NCEL) pelletizing test facility, located at the Naval Air Station in Jacksonville, Florida. During these tests, approximately 7 tons of RDF from Ames, Iowa and Pompano Beach, Florida were pelletized with the various binders. From this study, it was determined that binders increase the bulk specific gravity of the pellets substantially. The durability of the pellet is a direct function of the bulk specific gravity of the pellet. Therefore, the pellets made with binder were considerably more durable than those without binder. Additionally, the pellets produced with calcium hydroxide during this study were water resistant and storable with negligible chemical or biological degradation.

The next phase of testing involved a full-scale cofired combustion tests of binder enhanced dRDF pellets and high sulfur coal at Argonne National Laboratory (ANL) in Argonne, Illinois. These tests were conducted during June/July of 1987 in ANL'S spread-stoker, traveling grate type boiler, equipped with a lime sprayer and baghouse pollution control equipment. These tests were conducted with industry, state, and municipality participation both in the critiquing of the test plan and in witnessing of the actual test runs. Approximately 600 tons of densified RDF (dRDF), made in two locations, Thief River Falls, Minnesota, and Minneapolis, Minnesota, containing binder at levels of 0, 4 and 8 percent, were blended with high sulfur coal at ratios of 10, 20, 30 and 50 percent by Btu content. At each blend ratio, the response characteristics of the boiler were investigated. The ability of the boiler to gain or shed load with changing steam demand were investigated as well as the ability of the boiler to maintain a steady load. In addition to the blend ratio tests, identical evaluations were carried out when firing with 100 percent coal. These evaluations provided baseline data for comparison with the blend ratio. At each blend ratio the boiler was placed into steady state operation. During this study, 600 raw feedstock samples, through-grate/under-grate bottom ash samples, and economizer/multicyclone fly ash samples were collected. Additionally, 900 emission samples from the combustion chamber and the duct, both prior to and after the pollution control equipment, were collected. The samples were analyzed for chlorinated dioxins and furans, polyaromatic hydrocarbons, polychlorinated biphenyls, sulfur oxides, nitrogen oxides, chlorides, bromides, and trace metals. All ash samples from this pilot study passed EPA-toxicity tests, and emission data showed a decrease in organic gasses over those of 100% coal. Perhaps the most interesting result from this study was the reduction of sulfur dioxide and nitrogen oxide emissions over those of 100% coal. Sulfur and nitrogen oxide gasses emitted from the combustion of coal contribute to the formation of acid rain, which can harm aquatic life and damage houses and buildings. The test showed significant decreases in both sulfur dioxide and nitrogen oxides. When the coal was mixed with a 30 percent Btu RDF substitution, with a 4 percent binder content, sulfur dioxide emissions were reduced by 30 percent and nitrogen oxide emissions were reduced by up to 50 percent. This reduction in stack gas was due to the precipitation of the sulfur dioxide and nitrogen oxides in the form of calcium sulfate and calcium nitrate in the ash.

In order to aid in the commercialization of RDF as an alternative fuel source, potential markets were examined which could burn the RDF in existing facilities. The cement industry burns about 21 million tons of coal or its energy equivalent in natural gas or waste derived fuel each year, making it an ideal user of RDF. Our research team, through grants from the State of Texas, conducted a study at Texas Industries Inc. (TXI) in Midlothian, Texas, to demonstrate the technical, environmental, and economic feasibility of using RDF as a supplemental fuel in cement kilns. In October of 1990, approximately 160 tons of dRDF were shipped from Reuter Resource Recovery in Eden Prairie, Minnesota, to TXI cement plant in Midlothian. The dRDF was stored in a decommissioned oil tank to protect it from the elements. The dRDF was then "re-fluffed" in a grinder so that the particle size would be small enough to allow the material to be air-borne upon entering the cement kiln. The RDF was co-fired with coal and natural gas at rates of 15 to 20 percent by Btu content over a 65 hour period. Samples of coal, dRDF, kiln dust, and cement clinker were taken on an hourly basis. Additionally, raw feed (slurry) samples were taken every three hours. A total of 113 tons of the dRDF were burned over the course of the test. A key factor in producing quality portland cement is the consistency of the kiln combustion conditions. Therefore, feed end temperatures and chain section temperatures were also monitored. Results from the test indicated that kiln conditions could be maintained at a consistent level with the use of RDF as a supplemental fuel. Because the ash resulting from the combustion of the RDF actually ends up in the cement itself, contaminants such as heavy metals could cause deleterious effects on the cement. Therefore, the actual quality of the cement was also tested using standard ASTM test methods. Decreases in sulfur trioxide were observed owing to the fact that the RDF had a much lower sulfur content than the bituminous coal used by TXI. Also, an increase of over 35 percent was observed in the free lime content which was well within expansion test parameters for cement. Increased free lime is generally a measure of incomplete burning conditions in the kiln combustion zone, resulting in a decrease in tricalcium sulfate (C_3S). However, the

C_3S actually increased about 1 percent, and burning temperatures were slightly higher than the baseline. Further investigations showed that the RDF had a CaO content of almost 13 percent. Because the raw mix is designed to attain complete reaction of each component, the CaO contributed by the RDF was in excess, resulting in an increase in the free lime. To avoid this problem, the limestone in the original mix could simply be reduced. The typical physical cement parameters for normal consistency, setting time, flow, false set, fineness, and permeability of the cement produced during the test burn were not different from that of baseline cement. Furthermore, long-term compressive strength of the cement was not affected.

In order to determine the environmental feasibility of burning RDF as an alternate fuel in cement plants, emission studies were also conducted during the test burn. Sulfur dioxide and nitrogen oxide emissions were determined from an hourly monitoring system in the cement plant, while hydrochloric acid emissions as well as polychlorinated dioxin and furan analysis were contracted to a private lab. A decrease of 20 percent in sulfur dioxide and of 34 percent in nitrogen oxides were observed during the test burn. These reductions can be attributed to the lower amounts of sulfur and nitrogen in the RDF as well as the decreased availability of oxygen to combine with nitrogen due to the high fuel/air ratio. Also, a significant increase in hydrochloric acid was observed. This was attributable to the high amounts of chlorine in the RDF in the forms of plastic and inorganic salts. These HCl levels can cause corrosion to the equipment, especially to the refractory brick inside the kiln. For this reason, the HCl would have to be better controlled if RDF was used as an alternate fuel on a long-term basis. Two possible solutions to the HCl problems are: (1) addition of calcium hydroxide binder to the RDF to neutralize the HCl, and (2) stricter quality control of the RDF to remove more of the plastics. Note, however, that plastics contribute to the Btu content of the fuel. Therefore, an optimum level of plastic would maximize the Btu content, while keeping HCl levels within emission limits. The polychlorinated dioxin and furan levels were not different than those associated with burning 100% coal. Furthermore, the 2,3,7,8-isomer considered to be the most toxic was below detection limits. Because the materials entering the kiln (both in terms of the raw feed and the fuel) are finely ground solids, the kiln gases are laden with dust. Most of the dust (90-99%) is recovered by an electrostatic precipitator and landfilled. For this reason, analysis were also conducted on the kiln dust. The dust showed an increase in sulfur trioxide, suggesting that the SO_3 remained in the gas stream until precipitated out as kiln dust. Trace elemental analysis were not conducted during this test, but might prove interesting in future tests.

Finally, an economic feasibility study was performed to determine whether or not RDF is an economically viable substitute fuel for the cement industry. In general, benefit-cost analysis is a systematic tool used to help guide decision-makers by providing useful information about the possible effects of a given program. This study examined both the benefits (private and social) and the costs associated with co-firing RDF at Texas Industries cement plant under three potential alternatives including:

(1) Shipping RDF from the nearest available facility.
(2) Building a 200 ton per day (tpd) resource recovery plant on site and manufacturing the RDF for private use.
(3) Building a 400 tpd resource recovery plant on site and manufacturing the RDF for private use.

Some simple calculations can show the enormous potential benefits of co-firing RDF in cement plants. TXI currently produces approximately 1.2 million tons of cement clinker per year, burning about 300,000 tons of coal per year. Based on this study, RDF could be substituted for coal at 20 percent by Btu content, resulting in a coal savings of 60,000 tons per year. The state of Texas produces 9.7 million tons of cement clinker per year. Use of RDF state-wide could therefore result in a savings of over 485,000 tons of coal per year. Coal savings for the United States, producing 83.8 million tons of cement clinker per year, could be almost 4.2 million tons per year from the cement industry alone!

Measured private benefits for TXI include: income from tipping fees (the fees paid by garbage collectors to dump their garbage), income from recyclables removed at the resource recovery facility, and reduction of coal costs. Social benefits resulting from the substitution of RDF as an alternate fuel source are: reduction of landfills, conservation of our coal/gas resources on the national level, and decreased health, ecological, and aesthetic problems associated with acid rain emissions.

The costs of co-firing RDF with coal in a cement plant include the costs of constructing and operating a resource recovery facility, permitting, testing, and retrofitting of existing plants. Additionally, if a resource recovery site were not erected on site, the industry would also incur RDF fuel and shipping costs.

The benefit/cost ratio is simply the total benefits divided by the total costs. An alternative with a ratio greater than 1 is considered to be cost effective. The results from this study showed that shipping RDF from a facility approximately 100 miles away would not be economically feasible. However, both the 200 tpd and 400 tpd plants manufacturing RDF on site appeared to be quite cost effective, with the larger plant being the most attractive alternative.

Clearly the technology of converting municipal solid waste into an environmentally acceptable fuel source carries the potential of redefining our ideas about garbage. As a waste disposal solution, an alternate energy source, and a pollution control measure, refuse-derived fuel may very well be the commodity of the future.

NOTES

Correspondence should be sent to Kenneth Daugherty, Dept. of Chemistry, University of North Texas, Denton, TX 76203.

＊
＊＊

Ecology in Conservation
and Conversation

Neil Evernden

IF THE FIRST Earth Day did one thing, it was to make "ecology" a household word. Not that the word was used in the strict sense of scientific ecology—far from it.[1] Nevertheless, there was a sense of optimism, sometimes bordering on euphoria, that we had found our savior. Ecology, the neglected science, had been waiting in the wings for years, and was now summoned to center stage and directed to fulfill its destiny. The earth is a shambles: show us how to fix it.

Of course, this expectation was doomed to disappointment from the outset, since ecology as "pure" science is not particularly concerned with telling us how to do anything, but rather with trying to describe to us the way things are. Its means of doing so is the application of the assumptions of empirical science, based in turn on a theory of knowledge that assumes a fundamental division between the knower and the known, the subject and the surrounding objects. Yet even if ecology was not centrally designed to produce engineering solutions to environmental problems, it is nevertheless, as a consequence of its own intellectual heritage, able to embrace the notion that the world is ultimately knowable and manipulable. It is implicitly accepted that environmental engineering is possible, if only we arrive at a correct understanding of the way nature works. To put it another way, even though ecology is not in the forefront of technology, it is nevertheless *implicitly technological*. For, as George Grant noted, "technology is the ontology of the age." We see the world *through* technology, so to speak: technology is a mode of thought and a stance toward the world, not simply an application of sophisticated tools.[2] This, unfortunately, is something we generally overlook, for it is this predisposition which makes it seem obvious that certain actions, rather than others, are appropriate. In other words, we are limited by our prior agreement about the way the world is in our consideration of possible actions. In embracing technology, in the broadest sense of the word, we have bought a "package deal" which not only give us our means but our ends.

I raise this issue out of concern about the uses and expectations of ecology, and in the belief that some means must be found to overcome this limitation if we are to find any genuinely effective response to the ongoing "environmental crisis." If there is one role that the original Earth Day continues to play, it is to show us *that nothing has changed.* Earth Day 1970 gives us a bench mark by which we can evaluate our "progress": I would submit that a comparison of our current condition with that evident in 1970 would reveal that we have exacerbated rather than alleviated the situation. In part, that may simply be a consequence of the enormity of the problem or the so-called "supertanker effect." This refers to the fact that even if the Captain of a supertanker perceived a risk and attempted to turn his vessel around, it would take many miles to effect that maneuver. By analogy, even if we knew exactly what we needed to do to reverse the course of environmental decline, it would still take a very long time to accomplish it.

But even admitting that, I doubt very much that this is the basic reason for our apparent failure. Rather, I would suggest that we have not tried to turn our vessel at all: we continue with amazing confidence in our current course. Indeed, it is not simply our lack of significant change since the first Earth Day that suggests this; it is even more apparent when we look further back.

Let's consider the response of concerned people when the finiteness of America's wild forests began to be obvious. There was wide debate, with contrasting views expressed. One view, which has dominated ever since, was voiced by the forest manager, Gifford Pinchot. Pinchot acknowledged the threat to the future of the forests, and he prescribed professional, scientifically based action to resolve the problem. This action Pinchot called "conservation," which he described thus:

Conservation stands for the same kind of practical common-sense management of this country by the people that every businessman stands for in the handling of his own business. It believes in prudence and foresight instead of reckless blindness; it holds that resources now public property should not become the basis for oppressive private monopoly; and it demands the complete and orderly development of all our resources for the benefit of all the people, instead of the partial exploitation of them for the benefit of a few. It recognizes fully the right of the present generation to use what it needs and all it needs of the natural resources now available, but it recognizes equally our obligation to so use what we need that our descendants shall not be deprived of what they need.[3]

Notice the emphasis on resources for use, and for use now; on resources to be used for benefit of all (not just the privileged); on development proceeding; and on leaving some for future generations.

These are all familiar sentiments, and all figure prominently in contemporary pleas for environmental action. Consider the following example:

Humanity has the ability to make development sustainable—to ensure that it meets the needs of the present without compromising the ability of future generations to meet their own needs. The concept of sustainable development does not imply limits—not absolute limits but limitations imposed by the present state of technology and social organization on environmental resources and by the ability of the biosphere to absorb the effects of human activities. But technology and social organization can be both managed and improved to make way for a new era of economic growth. The Commission believes that widespread poverty is no longer inevitable.[4]

This is the modern iteration of Pinchot's conservation philosophy, as expressed by the Brundtland Commission. Notice the similarity: Pinchot demands use by all Americans, and the recent authors demand use by all humans. Also, they both insist that development must proceed, and that current generations can meet their aspirations while still leaving some for posterity. Both are entirely human-centered, both treat the world as just humans and resources, and both implore us to do more of the same (with better planning so the adverse effects are diminished, of course). Pinchot sums it up in saying "The first great fact about conservation is that it stands for development,"[5] while the recent report says that "Far from requiring a cessation of economic growth, it recognizes that the problems of poverty and underdevelopment cannot be solved unless we have a new era of growth in which developing countries play a large role and reap large benefits."[6]

In every case there is the assumption, tacit or explicit, that there is a single global niche, and it is ours. And since it is all "ours," if follows that seals eating "our" fish must be "controlled" as must wolves eating "our" deer. The emphasis on human-centeredness, and on the absolute necessity of control and manipulation of nature, is ubiquitous. But what is most interesting—and worrying—is that these two reports were written about eighty years apart, and yet *they both say the same thing.* The break-through of the much-touted Brundtland Commission is little more than Gifford Pinchot recycled for the nineties. There is nothing fundamentally new about it, and what is even more troublesome is this: that having tried the Pinchot experiment for the better part of a century, and having during that period sunk deeper and deeper into environmental despair, we *still* are not ready to concede that perhaps *the experiment has failed.* How much longer must it be tested before we conclude that it is not enough? Equally troubling is the possibility that we are incapable of doing anything but reinvent Pinchot every time we address the environmental crisis. Is this really all we can do? Or is it merely all we can think of doing, all that is conceivable with our modern tools of thought?

George Grant's suggestion that we have obtained a civilizational destiny by adopting the "package deal" of technology would seem to support this conclusion. That is, if technology is indeed the ontology of the age, if we can only see nature through the lenses of technological thought, then of course all we can consider is the same solution, over and over again: *we are given over to it*. It is our mind-cage.

Ecology, as a science based on the implicit expectation of prediction and control, fits all too well into the civilizational destiny, particularly in its "Imperial" form as described by Donald Worster in his seminal book, *Nature's Economy*. Moreover, it may be that we are all, ecologist and layperson alike, prevented from even entering into serious discussions of any differing assumptions simply because we all share this ontology and its attendant language. As Grant says, "The ontology expressed in such terms as 'the ascent of life', 'human beings making their own future', 'the progress of knowledge', or 'the necessity of interfering with nature for human good' could not be used against itself. But there is no other language available which does not seem to be the irrational refusal of the truths of scientific discovery."[7] And remember, by "scientific discovery" we essentially mean the revelation of truth: how could one possibly oppose that and retain any trace of credibility?

And yet this way of thinking has been challenged from time to time. And if conservation can be said to be an American invention, it was born with a non-identical twin. At the turn of the century, the two were embodied by Gifford Pinchot and John Muir. These were certainly not the only players, but they exemplify the contrasting positions so well that we often think of it as a two-man debate. Pinchot was intent on viewing nature as a storehouse of objects for human manipulation, while Muir tended to respect nature as an entity in its own right, independent of human purposes. I think it might be fair to say that he was concerned with *wildness*, which is of course just that which is not under the sway of human willing and purpose.[8] Of course, our current expectations make that seem a rather perverse sentiment, since *to ask what something is* is essentially *to ask what it is for*—the presumption of intervention is ubiquitous.[9] What we can *do* with nature—its "resource potential"—is more important than what it *is*, in all its wildness and immediacy.

However much we might admire the man, Muir's advocacy must surely seem impractical in the extreme. But of course, that is the point: "practical" is a synonym for doing something to, and for constantly doing more in the name of "progress." What Muir's position entails is *the concept of limits*, which is utterly incompatible with the assumptions of technology.[10] And yet, I would suggest, it is only the concept of limits which can save us. To some extent, that is apparent in the descriptions provided by ecology; the very surfeit of species is testimony to the role of limits in nature, each species being in effect a means to self-limitation which permits the avoidance of competition.[11] Were interspecies competition to be the rule, we would expect to have seen a victor by now. What we see instead are millions of "compromises," ways of existence which we call species and which permit the avoidance of direct competition with other species. In agreeing, so to speak, that "if you eat only the red berries, I'll eat only the green ones," the two creatures embody sets of limits that make them what they are. Each finds its "place" on the planet, and dwells within that place in security and in perpetuity. This is in sharp contrast to the "global vision" which sees the entire planet as the human niche, which inevitably puts us in competition with every other living being. To live is to trespass on human territory, and every seal in existence is guilty of eating "our" fish. The global vision is the guarantor of our claim to placelessness.

But of course they aren't "our" fish, and this is not "our" planet. It is the awe-inspiring sphere on which we make our home, but is not "ours" in any proprietary sense. To foster this phrase is to encourage the attitude of domination and to diminish the prospect for self-limitation. In the revelation of limits, as in the description of interrelatedness, ecology may indeed provide insight which could inform our conversation. But to this point, ecology has been allowed to contribute only to conservation, not to conversation.

I mean this in the very broadest sense, as in Michael Oakeshott's "conversation of mankind."[12] This conversation is "the meeting place of various modes of imagining" which one enters not to debate or to argue, and not to prove "true" one proposition rather than another. It is instead "an unrehearsed intellectual adventure" in which many voices are heard and the many "ways the world is" are engaged.[13] Each voice has its own idiom and expresses the world in its own way, so to speak. The more "ways" we entertain, the broader our acceptance of the experience of the world. But to assign a monopoly to only one or two voices is to accept and endorse the actions intrinsic to their idiom and to deny credibility to all others. Moreover, "for a conversation to be appropriated by one or two voices is an insidious vice because in the passage of time it takes on the appearance of virtue"—we become proud of having excluded all others and attending to only one or two honored voices.

> Consequently, an established monopoly will not only make it difficult for another voice to be heard, but it will also make it seem proper that it should not be heard: it is convicted of irrelevance. And there is no easy escape from this impasse. An excluded voice may take wing against the wind, but it will do so at the risk of turning the conversation into a dispute. Or it may gain a hearing by imitating the voices of the monopolists; but it will be a hearing for only a counterfeit utterance.[14]

There have been many counterfeit utterances since the original Earth Day. And to this point, "ecology" (in the broadest sense) has been permitted to speak only in the idiom of science. Even when its practitioners may have wished to speak differently, they were forced to "imitate the voice of the monopoly" in order to maintain credibility. The monopoly belongs to what Oakeshott calls *the voices of science and practical activity*, which speak of our principal preoccupations. But surely, if we have listened to those voices exclusively and have found ourselves sinking ever deeper into the "environmental crisis," we must conclude it is time to let others speak. That is, it must be time to enlarge the range of voices in our conversation, and with them the means of considering our relationship with the natural world. I cannot say which voices these may be, and I doubt that I could welcome all of them wholeheartedly. But it seems to me that we have no choice but to attempt this democratization of speech, and that we must at least consider the possibility that our two favored speakers cannot save us alone. And as a starting point I would suggest that we turn again to the other branch of American nature-reflection, not as a quaint topic or a historical oddity, but as a possible source of genuine insight. Pinchot has had his day; let's give Muir and Thoreau, and their intellectual descendants, our attention.

And in the process, perhaps we can engender a realm of *ecological studies* in which our dilemma can be discussed far more openly than is currently possible. Pinchot claimed that "The first duty of the human race is to control the earth it lives on."[15] There is some irony in this, since Rachel Carson, who might be said to have initiated the environmental movement with her book *Silent Spring*, said in her conclusions that "Control of nature is a phrase conceived in arrogance, born in the Neanderthal age of biology and philosophy, when it was supposed that nature exists for the convenience of man."[16] This was clearly a rejection of Pinchot's philosophy, yet it is the modern reincarnation of that philosophy which now claims to be the savior of the environment. The more things change, the more they stay the same. A historian once claimed that we tend to imagine the past and remember the future, meaning that we always explain the future in terms of what we already know—our predictions are almost always that there will be more of the same, some modified version of what we have already experienced.[17]

Oddly enough, any criticism of the standard pattern of speech is commonly interpreted as a recommendation that we "return" or "go back" to some previous cultural incarnation. And usually, such an assertion deals a mortal injury to the upstart critic, for everyone can see how impractical it would be to try to return to some pre-industrial utopia. Yet such a recommendation is seldom actually made, since it is abundantly clear that we cannot "go back" to anything. Why then does such an interpretation so commonly arise? Perhaps simply because our linear model of history contains only two poles, past and future, and if one criticizes the progressive trajectory toward the future, one must obviously be advocating its alternative, the past. But of course, it is the very idea of *the inevitability* of this trajectory that is a chief inhibitor of conversation: Grant's "civilizational destiny" again. There is no particular reason why we must picture our situation in this linear image, no reason that there cannot be detours or loops in the affairs of life. Nor indeed why "metamorphosis" could not be a more appealing image than the never ending ascent of the ladder of progress. It may be that the battle is to be fought at the level of our "ruling metaphors"[18] rather than at the level of politics and technology. One must hope, then, that ecological thought can contribute to the civilizational conversation in a different voice that is currently assigned it, and that the range of our "destinies" will be similarly enlarged.

NOTES

1. Not only was it used as a synonym for "natural environment," but also as something like "the search for knowledge of nature." In the strict sense ecology is the search for scientific knowledge of nature. For the most part, I will be using the term in the narrower sense, but will be advocating a broadening of its domain to include all means of "knowing nature."

2. George Grant, *Technology and Justice* (Toronto: Anansi, 1986), 32.

3. Gifford Pinchot, *The Fight for Conservation* (Seattle: University of Washington Press, 1967), 79–80.

4. Fro Harlem Brundtland, *Our Common Future/ World Commission on Environment and Development* (Oxford: Oxford University Press, 1987), 8.

5. Pinchot, *Fight for Conservation*, 42.

6. Brundtland, *Our Common Future*, 40.

7. Grant, *Technology and Justice*, 33.

8. See Tom Birch's discussion of wildness in "The Incarceration of Wildness: Wilderness Areas as Prisons," *Environmental Ethics* 12 (1990):3–26.

9. See Grant, *Technology and Justice*, 34.

10. The ubiquitous assumption that competition is "natural" and unavoidable—not to say desirable—is too apparent to require emphasis here. But there are consequences that even the most devout believer cannot avoid noticing. Unlimited competition surely means that we must be in direct competition with each other for all those—supposedly limited—resources. That realization is troubling for some, not for others. For those who are troubled by it, such as the authors of the Brundtland Report, the only solution appears to be the denial of scarcity. There are not real limits, only temporary ones. And with careful planning we can overcome that: it is a matter of central belief. This means that while we may be in competition, we can, in the long run, all win. And if we all win, and if there are only two things in the world, people and resources, then apparently the loser has to be "resources"—which means every living thing on the planet except us. The world is, then, one big niche, with seals audaciously eating "our" fish. The concept of limits is unthinkable, and the psychic investment we have in maintaining that state of affairs is enormous. Surely no religious zealot ever clung more tenaciously to a belief.

11. See Paul Colinvaux' interesting discussion of this in Chapter 13 of his *Why Big Fierce Animals Are Rare* (Princeton: Princeton University Press, 1979), 136–49.

12. Michael Oakshott, "The Voice of Poetry in the Conversation of Mankind," in *Rationalism in Politics* (New York: Basic Books, 1962).

13. Ibid, 198, 206.

14. Ibid, 202.

15. Pinchot, *Fight for Conservation*, 45.

16. Rachel Carson, *Silent Spring* (Boston: Houghton Mifflin Co., 1962), 297.

17. L. B. Namier, who said in his *Conflicts: Studies in Contemporary History* (London: Macmillan, 1942), 69–70, that "One would expect people to remember the past and to imagine the future. But in fact, when discoursing or writing about history, they imagine it in terms of their own experience, and when trying to gauge the future they cite supposed analogies from the past: till, by a double process of repetition, they imagine the past and remember the future."

18. The phrase was coined by Owen Barfield in his useful essay, "The Harp and the Camera" in *The Rediscovery of Meaning* (Middletown: Wesleyan University Press, 1977), 65–78.

*
**

Educating Environmental Scientists for the 21st Century

Kenneth L. Dickson
F. Andrew Schoolmaster
Samuel F. Atkinson

Introduction

OVER THESE LAST twenty years we have learned a great deal about the environment, its components and systems, and how it is being impacted by human activity. Concurrently, new technologies ranging from the personal computer to satellite surveillance have helped us to better understand environmental complexity. Unfortunately, we are still faced with daunting problems. Global climate change, deforestation, overpopulation, loss of biodiversity, and ways to establish sustainable development are but a few of the challenges we face. The need for environmental education has never been more acute. Nowhere is this more important than in higher education. As noted by Crowfoot (1990:9):

> Colleges and universities face the challenge of determining how they will use their resources to increase environmental literacy and educate new specialists. Every area of intellectual and professional life will be deeply effected by efforts to ameliorate environmental problems. . . . Each institution will need to choose carefully what it does in order to make the best use of its financial and human resources to address this critical problem.

Prior to the 1970s, interdisciplinary approaches to environmental issues were uncommon (Malone, 1990). Two events—the observance of the first Earth Day in 1970 and the enactment of the National Environmental Policy Act, which required the preparation of environmental impact statements—helped change this by drawing attention to the need for interdisciplinary approaches to environmental problem solving. Academia reacted to these events by creating a number of environmental studies/ science programs that were often part of the curricula in geography, biology or engineering departments and usually disciplinary or multidisciplinary at best, in their composition. Universities still struggle to implement truly interdisciplinary programs.

Stember (1991) provides a useful typology for defining disciplinary endeavors. At the base level is what is called an intradisciplinary endeavor where the focus is within the discipline. Crossdisciplinary refers to reviewing one discipline from the perspective of another. Several disciplines providing their particular perspectives on a topic constitutes multidisciplinary. Interdisciplinarity is more complex and results when the viewpoints of several disciplines are integrated to the extent that the individual identities of those disciplines are sacrificed to achieve a more holistic perspective on the topic in question. Stember's last category of study is transdisciplinary where the focus goes beyond the disciplinary perspective. In most cases, especially within colleges and universities, the disciplinary framework of departments and faculty identification with these departments dominates and transdisciplinary endeavors are difficult to realize.

Interdisciplinarity therefore is integrative in nature, draws from the perspective of several disciplines, and attempts to achieve a holistic understanding of the issue or problem under investigation. Environmental science is particularly well suited to an interdisciplinary approach on intellectual, practical and pedagogical grounds (Stember, 1991; Crowfoot, 1990). Intellectually, there is no single discipline or field of study that can provide all the theories, concepts, and methodologies required to investigate any except the most simple of environmental problems. For example, the study of global climate change, perhaps the most challenging and important of all environmental problems, is inherently interdisciplinary drawing from physical as well as social sciences and the humanities. Practically speaking, environmental problems do not lend themselves to disciplinary solutions. An examination of the total impacts of reservoir construction can only be addressed when biological and chemical as well as social, economic, and demographic information is synthesized and integrated. The pedagogical argument for an interdisciplinary curriculum featuring coherence and integration is well documented at the undergraduate levels (Cheney, 1989). A core curriculum featuring courses in the humanities, as well as social and natural sciences provides an interdisciplinary framework capable of facilitating the type of educational milieu advocated by Crowfoot (1990).

It is now widely recognized that an environmental science program should be interdisciplinary in nature. Unfortunately, effective implementation is not much easier now than twenty years ago when the first environmental studies/science programs began to be developed. It is our premise that universities must find ways to break down the barriers which hamper interdisciplinary programs if we are to produce environmental scientists capable of addressing the problems of the 21st century.

Recently, the opportunity arose at the University of North Texas to begin to address these challenges as we designed new M.S. and Ph.D. degrees in environmental science. This activity forced us to examine current curricula at our university, to become familiar with environmental science programs at other universities, to explore administrative alternatives for the degrees, and to rethink how to create an interdisciplinary atmosphere for an environmental science program. This made us realize that the new programs should not be modeled after programs which emerged in the 1970s and 1980s, that the skills needed by environmental scientists in the 21st century will be somewhat different than those used in the last twenty-five years and finally, that many of the historical barriers to interdisciplinary programs still exist and must be broached if universities are to meet the challenge.

Educational Goals of a 21st Century Environmental Science Program

THE BASIC GOAL an environmental science program is to produce graduates who have the ability to address current and future environmental problems. The global community has a staggering array of environmental issues to address. They range from cleanup of unique toxicants in soil and ground water to curbing global warming. It is obviously not possible to include specialized training on each environmental problem facing mankind in any environmental science program. However, it is possible to develop basic environmental problem solving skills which, when combined with the innate intellectual capabilities of students, allow problems to be addressed more effectively. It is our belief that the educational goals of a contemporary environmental science program should be to have a student:

- Develop an understanding of environmental resources.
- Possess a fundamental knowledge of engineering and ecotechnological approaches to addressing environmental problems.
- Know and be able to apply the principles of environmental resource management.
- Understand the collection, analysis and interpretation of environmental data.
- Have data management skills.
- Understand the fate and effects of contaminants and other stressors in the environment.
- Know environmental law, policy, regulations and institutions.
- Develop a strong environmental ethic.
- Be able to work effectively as an interdisciplinary team member.
- Conduct integrated risk assessments and be effective in risk communication.

Devising an educational experience that meets these educational goals is a challenging endeavor. Note that we did not call this a curriculum, as merely taking conventional courses will not allow these objectives to be met.

Environmental scientists of the 21st century will be required to have a fundamental understanding of environmental resources. They need to know how the abiotic and biotic aspects of the world interact. Strong physical science skills including hydrology, soils, geology, chemistry and physics are essential. However, equally important is an understanding of ecology and the social sciences. Formal training in resource management brings together the scientific and social dimensions of the environment. For example, water resources management involves not only water quantity and water quality, but water resources economics and water policy and law.

Historically, environmental science programs have emphasized engineering solutions to environmental problems. Environmental scientists of the 21st century will need ecotechnological skills to complement engineering skills. Ecotechnology is a relatively new field which attempts to utilize ecological principles to solve environmental problems. Examples are constructed wetlands for treating wastes, bioremediation of oil spills by stimulating naturally occurring microflora, composting of municipal waste water treatment sludges, restoration of mined lands, and organic agriculture. Future environmental scientists will need to learn how and when to apply these emerging ecotechnologies.

Knowing how to collect and analyze environmental data will be an essential skill. With the advances being made in automated environmental monitoring technologies, the environmental scientists of the 21st century will be overwhelmed by data. Likewise, it is highly likely that these scientists will be managing projects conducted by contractors who will gather and analyze data. Having a fundamental knowledge of the challenges of sampling strategies, familiarity with quality control and quality assurance techniques and strong data management skills will be essential. Use of remote sensing technologies to monitor regional, national and global quality of the environment will increase. Future environmental scientists will need survey research skills in remote sensing and geographic information systems.

The environmental scientist of the 21st century will use empirical and statistically derived mathematical models to assist in decision making about the fate and effects of contaminants. Students should be trained to be users of ecological models, environmental models, and regional and global scale models.

Whether the environmental scientist is in academia, industry or government, it is essential to understand environmental laws and regulations and the institutional responsibilities of local, state, national and international agencies. Current environmental science programs emphasize local, state and national laws and regulations. However, in the future environmental scientists must work in an international perspective if solutions to pressing problems are to be found.

Historically, environmental science programs have been focused on the physical and natural sciences and engineering. In the future the programs must incorporate as equal partners the humanities, social sciences and the arts. Environmental ethics should be an integral part of the curriculum. It is evident that classical technological solutions to the regional, national and global scale problems such as global climate change, loss of biodiversity and population growth are not likely. Instead, a combination of technological solutions and changes in social behavior will be needed. The environmental scientist of the 21st century must have a catholic understanding and perspective. This will require skills in working with interdisciplinary teams made up of colleagues from not only engineering and science but the humanities and arts.

Finally, environmental scientists of the 21st century must understand and be able to apply the techniques of human health and ecological risk assessment. Likewise, the scientist must understand risk perception and develop skills in risk communication. Risk assessment allows the allocation of time and effort to those activities which pose the greatest risk to people and ecological resources. To perform risk assessments requires an understanding of indicators of human and ecological health, comprehension of probability concepts, modelling skills and analytical thinking abilities. However, to be effectively used the environmental scientist of the 21st century must also be able to effectively communicate to fellow scientists and the public about risk. The success of our efforts to address the daunting environmental problems we face depends on effective communication of risk and informed decision making on the part of individuals. Thus, risk assessment skills as well as verbal and written communication skills are paramount for the environmental scientist of the 21st century.

With these educational goals in mind we set out to develop M.S. and Ph.D degrees in environmental science. Needless to say, to accomplish all of these goals would be beyond the scope of any program. However, we attempted to devise a program where basic competency related to the educational objectives could be developed by students pursuing the degrees. Naturally, the Ph.D degree program provides for the development of more in-depth skills.

The degree programs use a combination of conventional course work, individual research, thesis/dissertation research, studio courses, internships and a non-traditional faculty to accomplish the educational experience. The proposed curricula for the M.S. and Ph.D degrees incorporate these elements.

Program Design

THE COMPONENTS OF the environmental science program proposed at the University of North Texas are illustrated in Figure 1. An attempt to carefully analyze each component and its interrelationship with the educational objectives has been attempted. In the largest sense, without a complete melding of the components shown in Figure 1, the program will have only a minimal impact. Our goal is a complete interdisciplinary merging of curriculum, faculty, facilities and students.

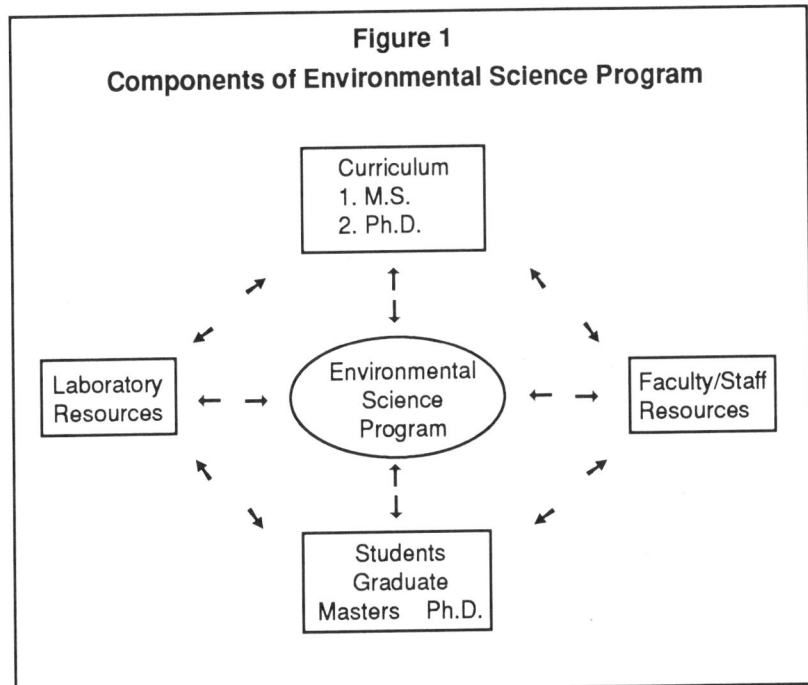

Figure 1
Components of Environmental Science Program

Table 1
ENVIRONMENTAL SCIENCE CORE - M.S. CURRICULUM

Biol 5010	(4)	Biostatistics
Biol 6340	(4)	Environmental Impact Assessment
Geog 5600	(3)	Seminar in Environmental Policy
Phil 5700	(3)	Seminar in Environmental Ethics

(One course from the following)

Biol 5030	(3)	Community and Systems Ecology
Biol 6360	(4)	Environmental Engineering
Chem 5390	(3)	Environmental Chemistry

17-18 Credit Hours

THESIS
(6 credit hours)
DESIGNATED ELECTIVES
(12 or more credit hours)

35-36 Total Credit Hours

Table 2
ENVIRONMENTAL SCIENCE CORE—Ph.D. CURRICULUM[1]

Biol 5010	(4)	Biostatistics
Biol 5030	(3)	Community and Systems Ecology
Biol 6340	(4)	Environmental Impact Assessment
Biol 6360	(4)	Environmental Engineering
Chem 5390	(3)	Environmental Chemistry
Geog 5600	(3)	Seminar in Environmental Policy
Phil 5450	(3)	Seminar in Philosophy of Ecology

27 Credit Hours

DISSERTATION
(12 credit hours)
AREA OF CONCENTRATION and DESIGNATED ELECTIVES
(51 credit hours)

90 Total Credit Hours

Curriculum

THE M.S. AND PH.D. core requirements are listed in Tables 1 and 2. The core for the M.S. program is composed of seventeen or eighteen semester credit hours depending on the five core courses selected. The courses were selected predominately from the existing inventory of courses and provide a foundation of natural sciences, social sciences and the humanities. This core assures that each student builds upon a common base facilitating a coherence in the early stages of their program.[2] The thesis option program requires twelve or thirteen semester credits of designated electives (minimum) which are selected in consultation with the student's graduate committee plus twelve hours of thesis work. If the student decides upon a non-thesis option, an additional eighteen or nineteen semester credit hours of designated electives are required, plus a final oral exam. To strengthen their methodological background, students will typically select courses in remote sensing geographic information systems (GIS), toxicology or chemistry as part of their elective requirements. Additional courses in limnology, water resources management, and natural resources economics have also been popular electives. Although few students have exercised this option, it is possible for cooperative education experiences (internships) to receive elective credit.

The Ph.D. degree curriculum is presented in Table 2. This curriculum represents the track taken immediately after a baccalaureate degree. Completion of a master's reduces the total hours to sixty. The interdisciplinary core of twenty-seven hours is complemented with twelve hours of dissertation, and fifty-one hours in the student's area of concentration with designated electives. The three areas of concentration available to Ph.D. students include land resources, water resources, and environmental management and policy. Individual departments contributing courses include biology, chemistry, math, computer science, political science, economics, geography and philosophy. Other subject areas of particular interest to a student may be used as elective hours with the approval of the student's graduate committee.

The goal of our program is to graduate students that are well versed conceptually and theoretically, and who are methodologically strong. The diversity of elective courses beyond the interdisciplinary core is intended to introduce students to the multidimensional nature of problem-solving and enable them to communicate with a wide range of environmental professionals. Students are required to identify and pursue research projects which provide the challenge of integrating information and applying research techniques. They are also encouraged to participate in conferences where they can present the results of their research and become familiar with the process of peer review. By providing three tracks of electives in addition to the core, we intend to produce graduates who are focused in at least one area while achieving an overview of the vast array of environmental issues.

Facilities

PROVIDING STUDENTS WITH the opportunity to acquire laboratory and methodological expertise is also an important component of the environmental science programs. Laboratory facilities available for student research projects are housed in the Departments of Biological Sciences, Chemistry and Geography. Additional off-campus facilities include a water research field station and aquatic resources laboratory.

Specific laboratory facilities available for students use are presented in Table 3. The sharing of these facilities by environmental science students requires a considerable amount of coordination and cooperation by the departments and faculty directly responsible for the individual facilities. This is especially true for equipment purchases and laboratory staffing.

Table 3
LABORATORY FACILITIES

Water Research Field Station
Remote Sensing–GIS Laboratory
Environmental Chemistry Laboratory
Aquatic Toxicology Laboratory
Geoarchaeology Research Laboratory
Cartographic Laboratory
Weather Station
Ecological Modelling Laboratory

Faculty

FACULTY FROM THE Departments of Biological Sciences, Institute of Applied Sciences, Chemistry, Geography and Philosophy are most directly involved in the environmental science program and are responsible for teaching the core courses. Faculty participation from the departments offering elective courses is coordinated by the environmental science advisors. Faculty from across the campus are eligible to serve on thesis and/or dissertation committees following Graduate School guidelines.

In order to expand on the faculty capabilities at UNT, a number of adjunct professorships have been established with environmental professionals in the region. Faculty privileges are extended to qualified individuals from federal agencies and industries. These adjuncts bring to the program new perspectives on environmental education and expand the role models to which students are exposed. Likewise, arrangements are made for UNT environmental sciences students to conduct research using facilities of the adjuncts. An example of such a successful partnership is the joint activities of the UNT environmental sciences program and the U. S. Army Corps of Engineers' Lewisville Aquatic Ecosystems Research Facility. Scientists with the Corps at the facility are involved in directing UNT environmental science students on research being conducted at the facility.

Staff members provide important support and instructional services. For example, the manager of the Remote Sensing and GIS Laboratory is responsible for the daily operations and scheduling of the use of laboratory equipment and performs research activities for external grants and contracts. The manager also plays an important instructional role in assisting students with remote sensing and GIS software such as IDRISI, ARC/ INFO, ERDAS, and GRASS.

Administration of the degree programs is through the Department of Biological Sciences' Division of Environmental Sciences. Chairpersons from each of the departments with core faculty have been very supportive and demonstrated considerable flexibility regarding faculty teaching loads, semester credit hour generation, space, equipment, scheduling of classes, and part-time teaching budgets.

Students

GRADUATE STUDENTS WITH a variety of undergraduate backgrounds are welcome in the program. Remedial work may be prescribed by the environmental science advisor where appropriate. Frequently students with undergraduate degrees in the humanities or social sciences are required to take leveling courses in mathematics and the natural sciences. Most students entering the M.S. programs have undergraduate degrees in biology, geography, earth science, or engineering. Most students entering the Ph.D. program have a master's degree in the biological sciences.

Students are encouraged to complete their core requirements quickly and to identify possible topics and problems acceptable for research projects leading to a thesis or dissertation. Course scheduling will be coordinated to accommodate traditional (full-time) and non-traditional (part-time, evening) students. Given the changing demographics of the graduate student population, and the university's functions as an urban, commuter-oriented campus, core courses will be available during the late afternoon, evening, and on the weekends to facilitate program completion. Financial aid in the form of research and teaching assistantships is available.

Overcoming Academic Barriers

HISTORICALLY, interdisciplinary programs have had to overcome a number of academic barriers to be successful. Environmental science programs are no exception. Strong academic discipline-oriented departments have been the basic organizational structure of universities for many centuries. Scholars in these departments have replicated themselves, thereby perpetuating disciplinary allegiances. Interdisciplinary programs such as environmental science have often been viewed as inferior in rigor by reductionists because they lack the depth perceived to be essential. Thus, environmental science programs have had to overcome the philosophical and psychological barriers caused by the academic department orientation prevalent at most universities.

The reward structure of universities also creates barriers for interdisciplinary programs. Most universities have reward structures that penalize faculty that participate in interdisciplinary activities.

Team efforts are frequently discounted, multiple-authored technical papers are given less weight for merit increases and tenure decisions, and collaboration on grants and contracts involving other departments are often discouraged. Why? Because of resource limitations in universities, departments are often competing with one another to get a bigger piece of the pie. Thus the catholic perspective is lost in an effort to demonstrate departmental productivity. Faculty often find that they are penalized financially and in regards to promotion if they invest their efforts in interdisciplinary endeavors. Feedback from the system discourages participation in multidisciplinary activities and interdisciplinary programs.

At the University of North Texas these problems have been partially addressed by administering the environmental sciences program through the Institute of Applied Sciences. By administering this program via an institute which has a clearly understood interdisciplinary research and graduate training mission, and strong administration support, it has been possible to work effectively with the academic departments. Key to the success of the program has been the concept that the Institute will staff the nucleus of the environmental sciences faculty, but that most of the faculty and courses will be taught in the traditional academic departments. This structure minimizes the threat to the academic departments that the environmental sciences will be a drain on scarce resources, insures that many of the course credits generated by environmental science students will be allocated to the academic departments, and establishes a reward mechanism for faculty participating in the program. The IAS is based on the "weak" center concept. It operates with only six faculty lines but has a strong administrative staff. The six faculty lines are filled with experienced senior faculty with a commitment to environmental sciences. They represent key disciplines such as environmental biology, environmental engineering, environmental chemistry, geology, archaeology and environmental law and policy. The strong administrative staff of the IAS provides grant and contract business management services, proposal and manuscript preparation to all participating faculty. The Director of the IAS has direct input to the academic departments and to the Deans on the performance of participating faculty. While this arrangement for administering the environmental sciences program at UNT does not fully overcome all of the academic barriers confronting interdisciplinary programs, it has proven to be an effective means of minimizing their impact on faculty and students.

Conclusions

DEVELOPING AN ACADEMIC program to produce environmental scientists capable of dealing with environmental problems in the 21st century is a challenging endeavor. First, the program content must impart skills which are applicable to problem solving over large geographical scales (local to global). Second, the scientists must be team players whose training is truly interdisciplinary incorporating physical and natural science with humanities, social science and arts. The program must develop a strong partnership with the social sciences, humanities and arts if it is to provide the essential atmosphere and training needed by environmental scientists. Finally, understanding of risk assessment, sensitivity to factors influencing risk perception on the part of the public and ability to communicate with a variety of publics are all essential capabilities that must be developed in the program. Special attention has to be given by the University and program administrators to overcome inherent barriers to interdisciplinary programs. Overcoming the barriers is essential if the program is to achieve the educational goals of equipping individuals with the skills needed to meet the challenges of the 21st century.

REFERENCES

Cheney, L. V. 1989. *50 Hours: A Core Curriculum for College Students.* National Endowment for the Humanities. Washington, D.C.

Crowfoot, J. E. 1990. Academia's Future in the Conservation Movement. *Renewable Resources Journal* 8(4):5–9.

Malone, C. R. 1990. Functional and Disciplinary Challenges for NAEP. *NAEP Newsletter* 15(5):4–5.

(NRC)National Research Council. 1991. Opportunities in Applied Environmental Research and Development. National Academy Press. Washington, D.C.

(SAB) Science Advisory Board. 1990. Reducing Risk: Setting Priorities and Strategies for Environmental Protection. SAB-EC-90-021. United States Environmental Protection Agency, Washington, D.C.

Stember, M. L. 1991. Advancing the Social Sciences Through the Interdisciplinary Enterprise. *The Social Science Journal* 28(1):12–14.

NOTES

1. The program described here is for a students entering the Ph.D. program after a Bachelor's degree. Students entering the Ph.D. program after a Master's degree will be required to take, or have already taken, the same core courses. If they have had some or all of the core courses in their master's program, they will substitute the same number of credit hours of designated electives in their degree program. All Ph.D. students will be required to distribute their designated electives between their declared concentration and their minor concentrations in the same proportion as indicated above.

2. The decision to use existing course work is largely attributed to the fiscal and administrative engineers of established new programs. Under the guidelines for new program development, new faculty lines, course proliferation, and increased expenditures for program delivery are to be minimized. This is especially true given the fiscal problems faced by most state-supported colleges and universities.

*
**

Part Three

Conservation, Economics, and the Corporation

*
**

The Corporate Responsibility
to the Environment

Jenny Cheek

Who would have thought twenty years ago that Earth Day would ever amount to anything? As a junior in high school, I collected trash from neighborhood streets, listened to rock bands in the park, wore bell bottom pants, and felt pretty darn cool about life. I also remember thinking that a guy named Jim sitting next to me was the cutest thing I had ever seen, and forgot completely the purpose of the Earth Day event.

The entire experience was actually lost on me until Earth Day 1990, when the environment was once again the hot topic of the moment, and people were asked to reflect on the changes made since Earth Day I. Public awareness and understanding have certainly increased, but only within the last couple of years has the corporate sector taken an active part in making responsible decisions where the environment is concerned.

One company's efforts to make a difference

DALLAS IS THE international headquarters for Mary Kay Cosmetics, the second largest direct sales company in the world. An in-house environmental and recycling program was developed in 1989, catching the attention and respect of not only Mary Kay's employees, but other major corporations as well. With the increased attention to the environment fostered by planning for Earth Day XX, one might think that the program was triggered by an awakening awareness of the earth around us. More realistically, one might think that operational economic incentives prompted development of the program. However, neither was the case.

Actually, a Mary Kay employee had a husband (mine) who didn't like to take out the trash. He delighted in seeing how many newspapers could be squeezed into the bedroom waste basket before it toppled over and someone else had to empty it. Surely there was something more worthwhile that could be done with all those papers! His example led to the idea for an in-house recycling program at Mary Kay Cosmetics. A proposal was soon formulated and presented to the CEO of the corporation. Being an avid environmentalist, he quickly approved the plan.

Two years ago corporate recycling was not a popular activity. In fact, it was pretty much unheard of. Yet a committee of enthusiastic employees took an idea of recycling newspapers and aluminum cans and expanded it to include office paper, plastic, glass, and corrugated cardboard. These commodities are now collected from Mary Kay's corporate offices, manufacturing facility, corporate warehouse, print shop, and local distribution center. The maintenance department picks up most of the items and assembles them in a central point, from which All Waste Paper Recycling hauls them away for sorting and recycling.

The recycling idea quickly caught on with Mary Kay's 1,500 Dallas employees, and as a result outdoor bins were provided so that they could bring these same items from home. People's recycling habits were changed as a result of the in-house program, not necessarily overnight, but changed nonetheless. It became important to be good citizens and recycle because it was the right and responsible thing to do.

Over five million pounds of recyclables have been eliminated from area landfills through the Mary Kay recycling program. There is a deep, underlying pride in the program because the employees had a voice in how it was designed. Through their active involvement and the encouragement from management, the program has been very successful.

The economics of the program were barely considered in the initial planning stages. Aluminum cans brought about two cents each at that time—estimated to generate approximately $2,000 annually through usage in the company's offices. Much to everyone's surprise, Mary Kay's in-house program generated $35,000 in revenues the first year and saved $30,000 in reduced waste hauling expenses. A large portion of these funds were donated to the Texas Nature Conservancy and The Dallas Parks

Foundation to promote positive environmental endeavors throughout the state.

Mary Kay was also instrumental in lobbying the Texas legislature to require the inclusion of SPI (Society of Plastics Industry) codes on all plastic containers. This makes for more efficient recycling by identifying specific resins in plastic containers. The code is included on the majority of Mary Kay's plastic bottles and jars.

All of Mary Kay corporate letterhead, business cards, order forms, package inserts, and most publications are now being printed on recycled paper. The cost of this paper is often more expensive than paper made from virgin pulp. Management, however, expressed a sincere commitment to purchase recycled materials whenever possible and has many times paid a premium to do so.

In 1990 the company began using soy-based inks for all in-house printing projects. Soy-based ink is non-toxic and makes the de-inking process easier. Paper recycling is eased considerably by alleviating the disposal of petroleum contaminants in the de-inking process.

Recycled paperboard packaging was also introduced in June 1990. To date, the company's orders of 2,577 tons of recycled paperboard have saved 43,809 trees, 18 million gallons of water, 10.5 million kilowatt hours of electricity, and 7,731 cubic yards of landfill space.

Mary Kay Cosmetics is the first major cosmetics company to introduce recycled packaging for such an extensive product line. Currently 90 percent of our products feature cartons made of recycled paperboard.

In addition, Mary Kay products are packed for shipping from its five distribution centers in recycled and recyclable corrugated cartons, using CFC-free polystyrene "peanuts" made from recycled resin. Because public perception of peanuts is sometimes negative, the company encourages its Sales Consultants to reuse these peanuts, donate them to mail centers, recycle them where available, or return them to the Distribution Centers for reuse. Other packing materials have been tested, including popcorn and shredded paper, but they did not properly protect the products from damage in shipment. Mary Kay will continue to investigate more environmentally friendly packing material as it becomes available.

In May 1990, Mary Kay Cosmetics received the Environmental Excellence Award from Clean Dallas, Inc. (a subsidiary of Keep America Beautiful). The award honored Mary Kay's successful employee recycling program, recycled packaging advances, and other environmentally responsible achievements.

The Corporate Recycling Council (CRC) of Dallas

MARY KAY COSMETICS is also proud to be one of the founding members of the Corporate Recycling Council of Dallas. The first organization of its kind in the nation, the CRC was formed to promote recycling in the business community through education and leadership.

In November 1989 the Texas General Land Office asked several business leaders to initiate a corporate recycling effort in Dallas (including Trammell Crow, *The Dallas Morning News*, U.S. Postal Service, and Dr Pepper and Coca-Cola Bottlers). The involvement of corporations, environmental groups, industry, and government agencies has raised the level of awareness concerning recycling's importance in reducing the solid waste stream.

In its continued effort to promote recycling, the CRC has offered three "how-to" seminars, with the last one offered in September 1991. More than 400 people have been taught how to implement in-house programs and have learned about key issues facing the public from legislative, industry and environmental experts. The CRC also encourages its members to support markets for buying products made from recycled materials. When materials like paper, plastic and glass are recycled into usable goods, a new market is created. For recycling to positively affect the environment, the market for recycled goods must be supported. In April 1990, the CRC presented the first Earth Friendly Awards, sponsored by the Mary Kay Foundation, to 46 Dallas companies that had the foresight to establish recycling programs prior to the current popular trend.

The CRC was honored with the Environmental Excellence Award in May 1991 for its outstanding assistance in promoting recycling awareness in the corporate community. CRC members have made a commitment to be environmentally responsible corporate citizens. More than 130 companies have joined the CRC since its inception. Following the lead of Dallas, other CRCs have been formed in Houston, San Antonio, Plano, and Longview, with two more being developed in Austin and Fort Worth.

The reality of recycling

THERE IS, HOWEVER, a very recent change being felt in the community regarding the ability to recycle logistically and financially. Because of increased public awareness and corporate involvement within the last year, the surplus of recycled materials has glutted recycling vendors. Currently there are not enough recycling mills and markets prepared to accept all the material being collected. As a result, the price for these commodities has declined substantially. A company may not now be paid the same high price for paper, plastic and glass, and may sometimes even have to pay a fee for pick-up service. It is also more difficult for smaller companies to recycle because the volume of their materials may not be enough for a recycling vendor to economically service the account. The CRC is developing a program whereby several small companies join together to recycle as one entity.

Although the ease of corporate recycling has diminished recently, it is expected to improve. As the public continues to ask for products made from recycled materials, new mills and markets will be developed.

Perhaps it's time to develop another new set of habits— habits that will focus on reducing the amount of materials and products we use, and reusing them whenever possible instead of automatically tossing them into the recycling bin. Several ideas include: the use of two-sided copies for memos and reports; writing rough drafts or informal notes on the unused side of otherwise discarded paper; personal coffee cups instead of disposable ones; and scrap paper turned into scratch pads. These ideas save not only resources, but money as well.

Sometimes being environmentally responsible can be tedious and time consuming, but it is always rewarding to know you are protecting just a little bit more of the world around you. It is also important to keep in perspective the real importance of the "3 R's"—because it's the right thing to do.

*
**

The Energy Business and Conservation

E. E. Spitler

THE WORLDWIDE energy business today is largely based on oil, coal, and natural gas. These are all fossil fuels which means they are non-renewable, at least on any time scale acceptable to the human species. It is a wonder of nature that dead vegetation trapped underground for millions of years has in selected areas been converted to these liquid, solid, and gaseous forms of stored energy.

The transition from renewable to non-renewable energy

OUR SPECIES HAS depended throughout most of its existence on renewable forms of energy, primarily wood and also foods to provide energy for human and animal labor. That first creative genius who learned how to control fire to keep warm or cook a meal undoubtedly used dead twigs and branches as fuel. Throughout most of our past existence, our energy needs, modest by today's standards, continued to be met by renewable resources (wood, dung, wind, water flowing downhill, labor from humans and domesticated animals). Only in the last few centuries have we learned how to extract large quantities of fossil fuels from the earth—first by mining coal and later by tapping oil and gas reserves thousands of feet below ground. Concurrently, technology was developed to make productive use of these new, suddenly abundant, forms of energy—steam engines, internal combustion engines, central heating systems for buildings, and, early in this century, powerplants to generate electricity.

This was what we labeled "The Industrial Revolution," and it had a profound influence on how we live and work in the industrial or so-called developed countries. Our energy usage has increased many fold as we replaced animal labor with relatively cheap fossil fuel energy. Other countries, which lacked access to fossil fuel resources or who moved less quickly in exploiting what they had, were left far behind in energy use, and the discrepancies between rich and poor regions of the world inevitably were magnified.

As in any human endeavor, the exploitation of fossil fuels has had unfortunate side-effects, some inevitable, others which could have been ameliorated if adequately foreseen. Human death and injury due to mining or processing accidents, environmental destruction from oil spills or drainage from coal mines, death and injury on the world's highways, and serious pollution of urban air are some of the social prices we pay for readily available, inexpensive fossil fuel energy. Yet, given all of these problems, on balance the people of industrialized countries have chosen to continue to increase their use of fossil fuels. The greater mobility, the more comfortable lifestyle, the ready availability of goods and services have seemed to be worth the cost. As demonstrated in recent years, we willingly go to war to assure a continued supply of inexpensive energy.

As long as people's needs and desires for fossil fuel energy continue, there will be an industry, an energy business, to supply the demand. But responsible leaders in that business recognize that the supply of fossil fuels cannot continue indefinitely, at least not at prices which could sustain continued growth in the business. Ever since the early 1970s fossil fuel energy prices, particularly oil and indirectly coal and gas, have related more to the ability of oil-rich countries to control and manipulate prices than to the actual costs of production. As easily exploitable fossil fuel reserves become limited to fewer and fewer countries in the future, this situation can only worsen. Increasing energy use by rapidly expanding populations in the developing countries will hasten the day when fossil fuels become expensive commodities.

Physically we are not going to run out of fossil fuels worldwide for hundreds of years. Tremendous coal, oil, shale, and natural gas resources remain. However, converting these resources to readily usable forms of energy, preferably liquid, and getting them to market will inevitably raise the cost of fossil fuel energy substantially.

Another cloud on the horizon for the energy business, speculated about for many years and now an issue of serious scientific concern, is the predicted global warming due in part to increasing use of fossil fuels. Even perfect combustion of a carbon-containing fuel can do nothing to prevent the accumulation of carbon dioxide in the atmosphere, and carbon dioxide is one of the chief suspects in creating a greenhouse effect. Granted, we are relying here on computer models with little hard physical evidence yet to confirm any warming effect. However, a scientific consensus has developed that warming will sooner or later occur to some unknown degree.

Faced with the fact that we are gradually using up low-cost non-renewable resources and the new concern about the effects of their use on global climate, all of us must consider how (1) to encourage development of alternative sources of energy and (2) to make more efficient use of our remaining fossil fuels. Consideration for future generations leaves us no other choice.

Some background

I RECENTLY RETIRED from the energy business after over thirty-one years with Chevron Corporation, a major international oil and gas company. My work was primarily on the technology side of the "downstream" business (refining, supply and distribution, and marketing). I had occasional contact with the "upstream" business (exploration and production). Over these years I have worked on and managed projects for solving environmental problems, improving energy conservation, and developing alternative fuels. I was also active in environmental organizations in the 1960s and early 1970s. In recent years I have served on the Editorial Advisory Board of the journal, *Environmental Ethics*.

Needless to say, I have strong opinions on the subject of the energy business and conservation. The remainder of this article will express some of these opinions along with illustrative personal experiences. The viewpoints are mine alone, although I doubt if they run seriously counter to those of the company that employed me.

Changing attitudes within the energy business

ATTITUDES TOWARD CONSERVATION within the energy business have changed substantially over the years. When I entered the oil business in the late 1950s, the emphasis was on maximizing the sales of products made from cheap crude oil. It was a highly competitive and growing business. Environmental problems were just beginning to be recognized, particularly air pollution in the Los Angeles Basin. Dr. Haagensmit at the California Institute of Technology had identified the motor vehicle as a major contributor to the Los Angeles problem, and the oil and automotive industries had cooperated in 1956 in making a first attempt at measuring emissions from cars. The test was conducted on the dry, paved riverbeds in Los Angeles, and has ever since been referred to as the Riverbed Test.

In the early 1960s a few of us at the research laboratory toyed with the highly optimistic idea of attempting to develop a "smog-free" car, but never believed we had a serious chance of selling the idea to management. The attitude of some members of the automobile industry was summed up by an engineering friend of mine who said that if the residents of Los Angeles had to spend one cold winter in Detroit, they would be much less concerned about L.A. smog.

Over the years these attitudes have changed. As the 1960s progressed, we all became more aware of mounting worldwide environmental problems. By 1970, the year of the first Earth Day, serious efforts were already underway to understand and attempt to solve these problems. Reflecting the public's strong interest in the environment, my Company introduced a new gasoline additive, called F 310, in January 1970. The additive removed deposits from critical parts of the intake systems of automobile engines, thereby reducing their emissions of harmful pollutants and restoring fuel economy. To promote the product, the Company developed television commercials illustrating the dramatic effect the additive had on the visual appearance of the exhaust from automobiles with very dirty engines.

The controversy resulting from the introduction of F 310 was interesting and instructive. As a researcher, I knew we had an outstanding new product which in widespread use would help improve air quality. I was dismayed when some other oil companies, feeling the competitive heat, claimed erroneously to have had equally effective additives all along. Further disturbing news was an announcement by the California Air Resources Board that it had found Chevron gasolines ineffective in reducing emissions in a very limited test of its employees' vehicles. Their initial state of dirtiness had not been determined. A final blow was a charge of false and misleading advertising by the Federal Trade Commission. This charge was reported nationwide by the news media.

Environmental groups were suspicious and probably resentful. I remember a meeting with the Sierra Club's 1970 President, Phillip Berry, in San Francisco. Chevron was represented by its Chairman, Bill Haynes; the Vice President for Research and Environmental Affairs, Eneas Kane; and myself. After I had made a slide presentation describing the additive and its effects as determined by extensive testing, there was somewhat formal and strained discussion. I got the distinct impression that the Sierra Club wanted us off its turf; in its eyes, the polluter was not supposed to take credit or gain business from reducing pollution.

Several years later the 9th Federal District Court of Appeals decided that, while the F 310 advertisements had the capacity to mislead, the additive in fact did all that we claimed. The finding received little news coverage. The Company had long since discontinued advertising the additive; but, to its credit, Chevron continued to blend it into its gasolines despite a substantial cost penalty relative to its competitors.

Early in 1971 I had the idea of bringing representatives of industry and environmental groups together for informal discussions. I was concerned about the widening gap and apparent misunderstandings between the private, for-profit organizations and the public, non-profit organizations. So I initiated a series of environmental luncheons in San Francisco in which environmental issues were discussed informally with no joint actions contemplated. Companies headquartered in San Francisco who were involved included Chevron, Southern Pacific, Del Monte, Pacific Gas and Electric, Utah International, and Bechtel. The Bay Area Council, a business association, was quite helpful in encouraging and participating in the luncheons. Environmental organizations included the Sierra Club, Friends of the Earth, environmental representatives from the League of

Women Voters, and People for Open Space. This last organization was a Bay Area group founded by a wonderful, dedicated lady, Dorothy Erskine. Dorothy attended the luncheons regularly despite her advanced age. I remember her expressing amazement that such a divergent group could sit down and eat together and discuss controversial issues in a civil and rational manner.

The luncheons continued on through most of the 1970s and, hopefully, moderated attitudes on both sides. They were a learning experience for me. I found there was much common ground between people working on environmental problems in industry and volunteers in environmental action groups. At the same time there were some unresolvable differences due to differing priorities. Too often, industry tended to be reactive— finding the most cost-effective way to meet an existing requirement. Environmental activists tended to not understand or to minimize the economic effects of their proposals. Nevertheless, I believe the luncheons were beneficial, and regret that I was unable to generate interest in similar environmental discussions in other metropolitan areas.

The 1970s saw major government initiatives at all levels in dealing with environmental problems. The oil embargoes and resulting gasoline shortages brought on by Middle East conflicts also pointed to the importance of energy conservation and the development of alternative sources. The energy business responded to these changing conditions, promoting energy conservation during periods of shortage in both 1974 and 1979. It also worked closely with government bodies, trying to assure that legal environmental requirements were both effective and practical.

Many of the laws had real teeth in them, including criminal penalties for corporate officers who failed to comply. I was asked to head a corporate task force on environmental compliance at Chevron in 1980 to determine what additional steps should be taken to assure full compliance with environmental, health, and safety regulations. After sampling both upstream and downstream operations, the task force recommended a number of actions. As a result, Chevron set up an environmental audit function at the corporate level, similar to the safety audits it had conducted for many years. Every operation was audited periodically for environmental compliance, a process which continues to this day.

Another recommendation by the task force which became Company policy was that all environmental and safety requirements must be met regardless of the degree of enforcement; and when a legal requirement was deemed inadequate, further precautions should be taken.

In recent years the energy business has shown more initiative in promoting sound legislation and regulations to help solve recognized environmental problems. Companies now accept the need for regulations which put all major players on the same competitive footing. A striking example of this approach was the joint proposal by oil companies and diesel engine manufacturers to limit the sulfur content of diesel fuels along with improving emission control systems in order to achieve the very low emission levels required by 1994. I was told that EPA officials expressed surprise that these two industries could agree on a common approach to such a controversial issue.

Another program initiated by Unocal, Chevron, Arco, and Mobil and which now involves fourteen oil companies and all three major U.S. car companies is aimed at modifying gasolines and further enhancing emission control systems to minimize the contribution of the automobile to urban air pollution. Such cooperative research programs have been encouraged by the U.S. Congress in legislation passed in 1984. The ongoing auto/oil research program should have an important influence on how the 1990 Clean Air Act requirements are implemented.

The origin of the program is worthy of some discussion. Oil companies clearly saw a business need to have lower emission gasolines along with better emission control systems compared with alternative fuels such as methanol, ethanol, or natural gas. These alternative fuels were being trumpeted by their proponents as "clean" fuels. The industry believed that improved gasolines accompanied by further evolution of engine control systems would be more cost-effective and certainly less disruptive in achieving air quality standards.

However, discussions between technical teams of the oil and auto industries, in which I participated at the early stages in the Fall of 1988, were not encouraging. The auto people resisted sharing their know-how on control of emissions with their competitors in a cooperative program. Emissions control has become an inherent part of overall vehicle powertrain design, and therefore a closely guarded secret in the highly competitive auto business. The Chairmen of the Boards of the various companies finally had to get together to resolve this dilemma and allow the research to proceed. Congress and the Administration also provided impetus by clearly signaling that further tightening of the Clean Air Act automotive requirements would be achieved in 1990 after several years of unsuccessful attempts.

Why have attitudes within the energy business toward conservation and the environment changed over the years? There are a number of reasons. First of all, legislative responses to environmental action groups at the Federal, state, and local levels have forced the energy business to face up to its responsibilities. As stated earlier, some of the laws include criminal penalties for noncompliance, and no members of top management want to risk going to jail. Also, legislation places similar requirements on all the companies within the energy business so that an individual company is not put at a cost disadvantage by being environmentally responsible.

Another reason attitudes have changed is the turnover of employees within the business, including top management. The new people come with environmental views reflecting those of the broader society, and as they move upward they have more and more influence on company policies and practices. A corporation is not a static entity. To survive and be healthy it must respond to the changing attitudes and values of its employees as well as its stockholders and customers.

Recently Chevron adopted a revised policy on safety, fire, health and the environment. Its goal "is to be a leader within industry by emphasizing innovation and encouraging creative solutions, both of which will improve our competitive position."[1] The policy states that the Company will conserve natural resources "by careful management of emissions and discharges and by eliminating unnecessary waste generation. This also includes wise use of energy in our operations."[2]

Ordinarily internal company operating policies are not put to a vote of stockholders. However, Chevron took this unusual step in 1991 with its environmental policy. Over 99% of the stockholders endorsed the policy, demonstrating that the owners of the Company overwhelmingly support sound conservation, safety, and health practices.

The message from customers of the energy business is less clear. Arco, followed by several other oil companies including Chevron, introduced reformulated gasolines during 1990 in urban areas of the country where air quality problems are severe. The purpose of the gasolines is to reduce hydrocarbon, carbon monoxide, and toxic emissions from automobiles. The Middle East crisis during the latter part of 1990 with its temporary large impact on gasoline prices probably influenced customer response to these new gasolines. In any event the response has not been overwhelming. While surveys indicate that people strongly support environmental improvement, they do not necessarily demonstrate their convictions by buying gasolines that pollute less.

Not being a market researcher, I hesitate to explain why. But I am concerned that a large credibility gap still exists between the energy business and the public despite the worthwhile changes in attitudes within the business. In particular, the oil business continues to have a major need to improve public understanding and acceptance. Companies spending millions of dollars to reduce the adverse effects of their products on the environment need to find a better way of convincing the public that the changes are truly beneficial and worth the incremental cost.

Distrust between the energy business and environmentalists

MANY ENVIRONMENTAL ORGANIZATIONS continue to see the energy business as an enemy. Some members of the business reciprocate the feeling. This mutual animosity is unfortunate because both parties have much in common. After all, all of us share the same environment and we benefit by minimizing waste. We should all understand that environmental goals can best be pursued in a healthy economy; or, to put it more bluntly, prosperous businesses are a prerequisite for real environmental progress.

On the other hand there are real and significant differences. Some conservation and environmental groups adopt philosophies that the energy business or any other business is unlikely to find acceptable. For example, the energy business survives by providing fuels and associated services to its customers at a profit to itself. Proposed actions which would seriously jeopardize or eliminate profits, block access to new sources of raw materials, or prevent construction of needed facilities will be fought vigorously by the business.

Within the oil business, probably the greatest concern is about blanket restrictions on exploration activities. Without new discoveries of petroleum the business is in trouble, particularly as more and more of the known reserves become concentrated in the politically unstable Middle East. Environmental organizations, rightfully concerned about the fragility of some of the areas proposed for exploration, should also understand the harm to the world's economy if no further exploration is allowed. And, to repeat, a weak economy is not conducive to protection of the environment. Ample evidence is available from the environmental records of the poor countries of the world.

A certain degree of skepticism over the motives and actions of the energy business is still understandable. But too much distrust leads to stalemate. I suspect that environmental groups would be surprised at the cooperativeness and genuine concern that the energy business would exhibit if allowed to proceed with future development plans in an environmentally responsible manner. There needs to be continuing, positive dialogue between key representatives of the energy business and environmental groups to understand each other's priorities and to seek common ground where possible.

The energy business and energy conservation

IN MY OPINION, the energy business has given somewhat mixed signals on energy conservation. While environmentalists seem to believe that energy conservation is the total answer to meeting future energy needs, the energy business puts more emphasis on finding and developing new energy sources. For sound business reasons, the industry is diligent in using energy carefully within its own operations. However, its efforts in urging its customers to practice energy conservation have been sporadic and usually limited to periods of national emergency. After all, more energy sales lead to greater returns, other things being equal. It is difficult for a business to continuously encourage less use of its products, even if they are largely based on non-renewable resources.

Here, I believe, the customer, the public, and the government have a role to play. Customers make their energy conservation decisions primarily by the types of equipment they purchase and how and to what extent they use the equipment. For example, a large, heavy automobile is generally more comfortable and safer than a small, lightweight vehicle. But it is also less energy efficient. When a customer opts for comfort and safety over fuel economy, the oil business is unlikely not to compete to sell all the fuel needed to meet the customer's requirements.

To turn this situation around, either government action or voluntary action by the customer is needed. The government can reduce fuel usage either by restricting the type of vehicle the customer can buy or by artificially raising the price of fuel through substantial increases in taxes. Of course, these government actions must be supported by the general public to be effective. An alternative, and to my mind a preferable approach, is for the customer to decide to sacrifice some comfort and safety as well as overall vehicle usage to achieve more efficient operation.

I believe the energy business as a responsible public entity should continuously encourage more efficient use of energy by its customers in addition to promoting development of new sources of energy. Some public utilities have found that emphasizing energy conservation among their customers is good business because it delays or negates the need for new, expensive facilities. This may be true as well for some other parts of the energy business.

The energy business and global warming

AS MENTIONED EARLIER, global warming has become a serious worldwide concern of the scientific community. Increasing usage of fossil fuels is one of they key culprits. How should the energy business, so dependent on fossil fuels, respond to this growing concern?

So far, the business has tended to point out the weaknesses of the scientific position. After all, even though the atmospheric buildup of carbon dioxide along with other gases of concern is an established fact, global warming itself is still a theory. The data are not conclusive. Scientists are relying on complex, but still crude, weather models to predict a warming effect. They could be wrong, but such an error is becoming more unlikely as research studies continue. The major scientific debate now centers on the extent of warming and its distribution over the earth.

My opinion is that the energy business must accept the scientific consensus and, painful as it may be, start promoting the concept of a gradual leveling off in fossil fuel usage world-wide. The potential consequences of continuously increased usage are simply too great. At the same time, efforts to decrease the usage of fossil fuels rapidly could be highly disruptive to the world's economy. I doubt if enough attention has been given to the risks of rapid cutbacks in fossil fuel usage. We need to know more about that side of the equation before we can make reasonable decisions. The energy business with its worldwide presence is in an excellent position to promote and support further research on the consequences (both economic and environmental) of various paths toward control of fossil fuel usage.

In the end, the governments of the major energy-using countries will have to play the lead role in choosing how much fossil fuel usage can be tolerated in the future. Only an international approach with broad societal support, including that of the energy business, can be effective.

The long-term future for energy

ASSUMING WE ARE successful in solving or at least ameliorating our energy-related problems over the next fifty or so years, what might the long-term energy future look like? Surprisingly, we need to worry about this question now because major changes in the energy business require many years to implement. We have a moral responsibility not to impoverish future generations by using up currently accessible forms of energy without providing for their replacement.

I suspect that we will eventually move from chemical forms of energy to nuclear. The invention of controlled fire, that is, energy released from combustion of organic materials, has reached its ultimate expression with the rapid exploitation of fossil fuels. We must now begin the transition toward the energy locked in the nucleus of the atom.

My crystal ball is cloudy when it comes to predicting the optimum approach toward nuclear energy. Solar energy is an approach widely favored by environmental groups. Having managed a solar energy research project for several years, I am painfully aware of its major drawback—its diffuse nature. Having the sun carrying out a nuclear reaction about 93 million miles from the earth has worked out well for the evolution and continued energy support of biological life, including pre-industrial humans. However, huge investments and extensive land use become necessary if we try to collect solar energy in sufficient quantities to carry out the activities of modern human life. For example, much of Arizona and New Mexico would have to be paved with photovoltaic cells to supply the current U.S. demand for electricity. Certainly, solar energy research should continue, but a nuclear powerplant that far away may not be practical to meet most of our needs unless we radically change our lifestyles.

So what forms are practical? I doubt if we know yet, so further work must be done on all reasonable possibilities. Fission is frightening; fusion is immensely difficult. But we should not be overly intimidated by the difficulties, nor should we dismiss reasonable choices because of overwrought, emotional reactions. Some risks with new energy sources are inevitable just as continued reliance on fossil fuels holds risks, political as well as environmental.

I believe the energy business should invest effort in seeking out the best alternatives to fossil fuels for the long-term future. Such an investment cannot be justified by traditional economic analysis because of the long time horizon. Traditional analysis substantially discounts any returns beyond ten years. But the investment could be justified on grounds of maintaining good public relations and meeting a moral responsibility. A responsible business should look after the needs of future generations when it is using up an irreplaceable resource to meet the needs of present generations. In addition, if the use of fossil fuels is legally restricted worldwide, the energy business must find alternatives if it is to continue to grow.

In the U.S. an alternative energy research and development program should involve all major players in the business—oil, coal, and gas producers and suppliers as well as the electric utility industry. A cooperative effort would be appropriate because patent protection is meaningless in such a long-term program. The industry-sponsored technology effort would complement basic and applied research conducted or sponsored by governments. Governments could provide incentives to an industry effort by not taxing earnings devoted to long-term alternative energy research and development.

Conclusion

In my view the energy business has come a long way in its attitudes and its actions on conservation matters, but it has further to go. It needs to promote efficient use of its products on a continuing basis. It needs to recognize the real likelihood of global warming and support a leveling off of worldwide fossil fuel usage. And it needs to conduct research on long-term energy sources which will allow future generations to enjoy a level of energy usage similar to our own if they choose to do so. These actions might admittedly make the energy business slightly less profitable at present. However, they could also assure its long-term survival and improve its acceptance by the public, the governments of the world, and hopefully even environmental organizations.

Notes

1. Chevron Corporation Proxy Statement, March 19, 1991, Exhibit A-1, Policy 530—Safety, Fire, Health, and Environment.

2. Ibid.

*
**

Economics and Resource Conservation

Michael L. Nieswiadomy

THE ENVIRONMENTAL PROBLEMS that society faces today are well publicized in the popular press. The greenhouse effect, depletion of the ozone layer, deforestation, water shortages, air and water pollution, trash on beaches, extinction of animal species, and overburdened landfills are a few of the problems that have received attention. However, potential economic solutions have not drawn as much attention, but are urgently needed because the public and scientific community must carefully weigh the cost and benefits of our actions. Although applying economic tools to solve environmental problems may appear to be a strange concept, environmental and natural resource problems are fundamentally economic problems, because they result from the economic decisions of individuals. The fundamental issue faced in natural resource economics is how to allocate scarce resources among competing ends (e.g., competing uses for a river). Most environmental problems occur because an agent is not required to bear the full burden of the costs of an activity.

Let me begin by providing a simple definition of natural resource and environmental economics: the study of the efficient (i.e., prudent) use of our natural resources and the environment in both the present and the future. Although this definition is apparently simple, the meaning of the term "efficient" requires further explanation. To understand the meaning of efficiency one must first recognize that all actions involve benefits and costs. Secondly, let us agree that we want to maximize society's net welfare (i.e., satisfaction), which we will define as difference between the benefits and costs. Therefore, efficiency requires that any activity (e.g., producing or consuming a product or service) should be carried out up until the point where the additional (i.e., marginal) benefit is equal to the additional (i.e., marginal) cost. If the additional benefit is greater than the additional cost, then activity should be increased, thereby increasing society's net welfare. On the other hand, if the additional cost is greater than the additional benefit, then the activity should be reduced. The concept is not really complicated; it is one we practice in our lives everyday. For example, if we buy five apples at 50 cents each at the grocery store, we purchased the fifth apple because it gave us at least 50 cents worth of satisfaction. We did not buy the sixth one, because it did not give us at least 50 cents worth of satisfaction.

Unfortunately, when we consider environmental issues, we often forget to use the same analysis. For example, in my environmental economics class I often ask the question, what is the efficient level of pollution? Quite understandably, many students respond that the efficient level of pollution is obviously zero since any pollution is undesirable. Of course, when I present the hypothetical costs of removing (or controlling) 100% of the pollution, no one is willing to pay. Evidently, we are willing to tolerate some level of pollution. Once a student has been presented this basic argumentation, the above definition of efficiency seems reasonable. The really difficult issue becomes one of determining the appropriate measures of benefits and costs. More will be said on this later.

The purpose of this paper is to familiarize non-economists with some of the solutions proposed by environmental economics. This paper is organized in six sections as follows: the concept of scarcity of natural resources; common property problems; the concept of external environmental costs and the roles of the market and the government in solving environmental problems; governmental mismanagement of natural resources; issues in measuring the value of environmental amenities (e.g., wildlife); and innovative economic solutions to disposal of waste and water shortages.

Natural resource scarcity

MANY OF US ARE naturally concerned about the potential for running out of natural resources such as oil, metals, and water.

Doomsayers gloomily preach that we are running out of everything. Economists take a more optimistic view of the world
based on society's ability to deal with many past so-called
resource crises. In 1980 an ecologist and an economist bet
$1,000 on the future price of five metals. (*Reader's Digest*, 61).
The ecologist, Paul Ehrlich, of Stanford University, has received
much attention due to his book, *The Population Bomb*, published
in 1968. In 1974 he predicted that "before 1985 mankind will
enter an age of scarcity" in which "the accessible supplies of
many key minerals will be nearing depletion." The economist,
Julian Simon, of the University of Maryland, has not received as
much attention. Simon challenged Ehrlich to select any natural
resource—grain, oil, coal, timber, metals—and any future date.
In October 1980 they bet $1,000 on five metals—chrome, cop-
per, nickel, tin and tungsten. If the 1990 inflation adjusted price
were higher than $1,000 Simon would pay the difference but if
the price were lower, Ehrlich would pay. Last fall Ehrlich wrote
a check to Simon for $576.07. Each of the inflation adjusted
prices had fallen. Why did Simon win the bet? Economists
believe that entrepreneurship and technological improvements
allowed these metals to be mined at a lower cost. In fact
economists' main argument with ecologists is that mankind has
the ability to adjust to scarcity. In fact, scarcity, in driving up
the prices of resources, plants the seeds for its own solution.
Individuals have greater incentives to find solutions to the
scarcity problems because the rewards will be greater. However, it should be noted that government interference can exac-
erbate the problems. If for example, the government artificially
holds down the price of a scarce resource, the wrong signals are
given to the market.

History has provided us with many lessons of man adapting
to scarcity. The Greeks switched from bronze to iron due to a
reduction in trade of tin. Of course, it was by experimentation
that iron was developed, but the catalyst for the experiments was
the reduced availability of tin. Similarly, the shortage of timber
led many to predict that the manufacturing industry would leave
England for Germany (where timber was more abundant) in the
16th century, but then coal was discovered. Many people are not
aware that the first energy crisis in the U.S. occurred in the
1840s and 1850s, when whale oil was in short supply. But

petroleum was discovered and no one thinks about the whale oil crisis. In the 1960s and 1970s in the High Plains of the U.S. many were predicting the drying up of the aquifers (in particular the Ogallala) and dire consequences for agriculture. But more efficient irrigation practices have postponed this "crisis" (Maurice and Smithson).

Common property problems

ALTHOUGH ECONOMISTS ARE generally optimistic about the market place solving resource scarcity, they agree that the problem of common property needs to be addressed. Common property resources are those not exclusively controlled by a single agent or source (Tietenberg, p. 47). When our legal system does not establish property rights, numerous examples of common property problems exist such as wildlife, air and water. The primary problem facing common property resources is the "use it or lose it" incentive. For example, the American bison was nearly made extinct. Since no one owned the buffalo, each hunter had an incentive to kill a buffalo before someone else did. No one had an incentive to preserve the herd. Oil fields and aquifers face similar problems.

The African elephant is the modern day equivalent of the 19th century American bison. The plight of the elephant is well known, as their numbers have diminished rapidly. Two different solutions have been attempted. One method, limitations on ivory exports, has been implemented in Kenya, Tanzania, Gambia and Somalia, yet the elephant population has continued to decline. The other approach, known as the property rights approach, has been utilized in Botswana, Namibia, South Africa and Zimbabwe. The elephant population has risen in the last decade in these countries. In Botswana, for example, it has risen from 20,000 to 50,000 today. Permits to hunt elephants are issued to villages, and they can be bought and sold. A village that increases herd size is given greater permits, thus it has incentive to preserve it herds.

The solution to saving the elephant is to understand that the problem is an economic one. The Convention on International Trade in Endangered Species of Wild Flora and Fauna (CITES) ban on ivory exports, sought by World Wildlife Fund and other conservation groups, although well intentioned, has not been successful. The solution lies in establishing property rights for the elephants. Of course the establishing of property rights requires some government intervention, but the market place should be allowed to function once the property rights are established. Strict government bans are not likely to work. Although this concept cannot be applied to all forms of wildlife, it deserves serious consideration from all environmentalists. Good intentions alone are not enough.

Environmental externalities

ENVIRONMENTAL PROBLEMS such as pollution are also common property problems. For example, individuals have historically dumped wastes into nearby rivers because no one owned the river. However, since environmental problems pose a significantly different set of issues, we often treat them separately. Most environmental problems occur because an agent is not required to bear the full burden of the costs of an activity. For example, if a steel mill owned a resort hotel immediately downstream from its plant, it will bear the burden of dumping its pollution, and it is unlikely to pollute. However, when the firm does not bear the full costs of its actions we say that a negative externality exists. When negative externalities exist, too much of the good (e.g., steel) will be produced at too low a price.

The focus of environmental economics is to provide solutions to environmental problems that incorporate environmental costs into the decision making process, but at the same time *minimize* the cost of implementing these solutions. If these objectives are met, society's desire for consumer goods can be balanced with its desire for a cleaner and safer environment. This is another area where economists recognize the need for governmental intervention, but the market place should be used to provide the proper incentives, as explained below.

At times there may be some misunderstanding about what environmental economists mean by "attaining a goal at a *minimum* cost." We do not mean that we will lower the qualitative standards of the goal (i.e., perform shoddy workmanship) in order to lower the costs. We will find the least cost method of attaining the goal, precisely as it is defined, both quantitatively and qualitatively. For example, if the government establishes tradeable emission permits, firms that can reduce emissions at a lower cost will sell their permits to firms that have high costs of reducing emissions. This system is far superior to a "command and control" approach that mandates the emission reductions for all parties. The tradeable permit system can achieve the same emissions reduction goal at a lower cost. Furthermore, firms have incentives to reduce pollution control costs because they can sell even more permits to other firms. There are many other innovative methods proposed by environmental economics, far too many to mention here. (See Tietenberg for further discussion.)

Government mismanagement of our natural resources

IN THE PREVIOUS sections we have shown that there is a need for government intervention when common property and negative externality problems exist. However, environmental economists have also shown that the government has mismanaged some national resources. Two examples are examined in detail by Baden: the National Park Service's destruction in Yellowstone and the Forest Service's excessive timber cutting program.

Yellowstone and the National Park Service. The argument that the government has destroyed much of Yellowstone is espoused by Alston Chase, the former chairman of the Yellowstone Library and Museum Association (the official publisher of books on the natural history of the park). He states:

> Over the last seventy years nearly every conceivable mistake that could be made in wildlife management has been made by the Park Service in Yellowstone. Not a year has gone by since it assumed responsibility there when the National Park Service . . . did not kill an animal in the name of an environmental ideal. Today its management policies threaten the very capacity of the park to sustain life. (Chase, p. 233)

Wildlife management experts have long recognized the important ecological relationships among predators and prey. But Yellowstone's managers have often interfered with this delicate balance, with devastating results. In the first twenty-five years of the park's history, government managers killed 4,889 coyotes, 781 mountain lions, 554 bobcats and 20 wolves (all predators of the deer), with disastrous consequences. As the deer population expanded, there was insufficient food left to support the population. But the problems extended to other aspects of the park's life as well. The elk population also grew in the absence of its natural predators. The elk ate the young shoots of willow and aspen, leaving fewer trees for the beaver to build dams. Since the beavers could not build their dams, erosion increased and the population of trout (which need clear water to lay their eggs) decreased. Normally the beaver move into a stream lined with willow and aspen and eat the bark and use the trees to build their houses and dams. After they have exhausted most of the available food, they move on. However, their dams raise the water table and foster new willows and aspen. The beaver return in twenty or thirty years and the cycle starts over. However, this cycle was irrevocably destroyed by the large population of elk.

The managers of Yellowstone have caused many other problems as well (see Chase). But how have they been able to do this? First of all we must understand that any bureaucracy must seek political allies to protect and expand its budget. The Park Service first used the railroads and later used environmental groups and park visitors as its allies. But the Park Service has not allowed any independent research to be conducted, so the public does not really know the impact of its policies. It fact, the Park Service spends less than two percent of its billion dollar budget on research, and it has only twenty-five natural scientists with Ph.D. degrees.

Fortunately, there are alternatives to governmental control of our parks. The Audubon Society runs the 26,800 acre marshland Rainey Wildlife Sanctuary in Louisiana quite successfully. The Nature Conservancy owns and manages a national system of nearly 800 sanctuaries. It has been very successful, preserving about 2.4 million acres of land since 1951.

The U.S. Forest Service. The U.S. Forest Service has a track record as bad the Park Service. According to Baden, the Forest

Service has violated its custodial responsibility by serving special interests whose goals conflict with environmental protection. It may be surprising to learn that the Forest Service has directed the construction of over 342,000 miles of roads, which is eight times more than the total mileage of the U.S. Highway Interstate System. Along with its road building program the Forest Service has encouraged subsidized timber sales in areas that are ecologically sensitive and economically unprofitable for a private firm to harvest without subsidies. These programs have caused serious erosion problems in high mountainous regions.

Why has this travesty occurred? The answer lies in the incentive structure faced by the Forest Service. Because of the Knutson-Vanderberg (K-V) Act of 1930, the Forest Service's budget is linked to number of board feet harvested. The Forest Service is able to increase its budget by encouraging greater harvesting of timber, even to the detriment of the environment. Of course there are many political interest groups that benefit from these policies. Thus, it will take tremendous political pressure from environmental groups to change this setup. Until then, the environment will continue to suffer.

Valuing environmental amenities

ONE OF THE unique contributions made by environmental economists in the past few decades is to provide some tangible ways to measure the value of environmental intangibles such as beautiful rivers, mountains and wildlife. Since many of these goods are not, or cannot be sold in the market, no market prices exist to measure their value. But first we need to define the different ways economists value wildlife and other environmental amenities. There are usually considered to be two types of use values (Eubanks): consumptive and nonconsumptive. Consumptive uses, such as hunting and fishing, diminish the wildlife population. Nonconsumption uses, such as photography, do not diminish the population. Environmental economists have been analyzing these issues for some time. However, more recently, attention has been given to three categories of intrinsic values: existence value—where people value the existence of a wildlife resource, even if they do not expect to use it directly; option value—where people want to retain the option of future use; and bequest value—where people want to assure the existence of wildlife populations to future generations.

Two models are often used to estimate these values: contingent valuation models and travel cost models. In a contingent valuation model, people are asked to answer survey questions regarding their values. The travel cost model is most often used for hunting, fishing and recreational activities. Travel costs, including the cost of time, represent the price "paid" to receive the recreational benefits. Using statistical regression techniques, a demand curve can be constructed. This method is often referred to as the Clawson-Knetch method (Clawson and Knetch).

Innovative economic solutions to environmental problems:

Two brief examples of managing water shortages and waste surpluses can be given to illustrate how economic tools can be utilized to solve environmental problems.

Water shortages. In the area of water use, environmental economists can devise water rate schedules that charge efficient prices for the water. The efficient price for a marginal unit of water should be equal to the true marginal cost of providing the water. For example, in Texas, the price of water should be higher in the summer than in the winter.

Waste surpluses. One solution to the problem of illegal dumping is to implement a "deposit-refund system" like those put into effect by the Federal Republic of Germany for oil (Page), and by Sweden for abandoned cars (Bohm). In general, the system would function as follows:

(1) Using all research to which it had access, the government would determine those biological, chemical, and radionuclide substances which had the potential to contaminate the environment, e.g, groundwater.

(2) Based on the substance's potential for damage to the environment, the government would assign a "deposit" which must be paid for every unit of the product that is produced in or imported into the state.

(3) The deposit money would be held in a special trust fund. It is refunded when proof is provided that, after use, the substance

was disposed of in a manner which was environmentally sound. The refund is a direct refund if the original producer or importer takes care of the final disposal or is an indirect refund in the form of "free disposal" if another party takes care of the final disposal and collects the refund.

(4) If the substance is disposed of improperly, monies from the trust fund would be available to repair the damage to the environment.

This type of program has one key advantage over the currently used national Superfund program—the deposit-refund system is proactive rather than reactive like the Superfund. Since the disposal costs will be paid up front, users of the final product will have little incentive to dispose of the consumed product improperly—a phone call will bring someone to pick it up. However, if this free disposal option is not available (as is presently the case) many individuals and firms will be tempted to dispose of the consumed product illegally to reduce their disposal costs, especially when budgets are tight. Furthermore, the deposit-refund system also encompasses the present feature of the Superfund program—cleaning up existing contamination sites. One caveat must be mentioned. The deposit-refund system can encounter high administrative costs, so care must be taken to select only those contaminants for which the benefits of the system clearly exceed the administrative costs. Rigorous economic analysis is needed.

Such a system incorporates the best features of both the market system and of governmental controls. Namely, the level of protection desired by the citizens can be achieved and, at the same time, the price system provides incentives for everyone to dispose of wastes safely.

Conclusions

THE FIELD OF environmental economics is a dynamic discipline offering solutions to a wide array of environmental problems. Hopefully this brief paper has conveyed that message. Some of the proposed solutions may seem strange at first glance. But if we wish to deal with the vast numbers of environmental problems facing us today, we must avail ourselves of all of the tools possible.

In summary, I request the environmentalist to be wary of proposals offering the socially acceptable options. In this era of "After Earth Day," environmental economics teaches us to look out for the hidden consequences of so-called "solutions." Good intentions alone will not solve our environmental problems. Environmental economics offers a methodology to prudently manage our natural resource and environmental problems.

REFERENCES

Baden, John, "Destroying the Environment: Government Mismanagement of Our Natural Resources," National Center for Policy Analysis, Policy Report #124, October, 1986.

Barton, Denise, "Conservationists vs. the Elephant," *The Margin*, January/February 1990, 24–25.

Bohm, Peter, *Deposit-Refund System: Theory and Application to Environmental, Conservation, and Consumer Policy*, (Baltimore, Md: Johns Hopkins University Press for Resources for the Future, Inc., 1981).

Chase, Alston, *Playing God in Yellowstone: The Destruction of America's First National Park* (New York City: The Atlantic Monthly Press, 1986).

Eubanks, Larry, "How Do We Value Wildlife," *The Margin*, September/October 1990, 16.

Maurice, S. Charles, and Charles Smithson, *The Doomsday Myth: 10,000 Years of Economic Crisis* (Stanford University: Hoover Institute Press, 1987).

Page, Talbot, *Conservation and Economic Efficiency: An Approach to Materials Policy* (Baltimore, Md: Johns Hopkins University Press for Resources for the Future, Inc., 1977).

Tierney, John, "A Bet on Planet Earth," *Reader's Digest*, March 1991, 61–64, from *New York Times Magazine*, Dec. 2, 1990.

Tietenberg, Tom, *Environmental and Natural Resource Economics*, Second Edition, (Glenview, Illinois: Scott, Foresman & Co., 1988).

＊
＊＊

PART FOUR

ENVIRONMENTAL PHILOSOPHY

*
**

Deep Ecology and Nonhuman Others

Dolores LaChapelle

Man is not the supreme triumph of Nature but rather an element in a supreme activity called Life. —R. Murray Schafer

THAT FIRST EARTH DAY, back in 1970, it seemed so simple. We would merely pass laws and regulations to clean up the rivers and the air, and to control wastes—make an environment more fit for human life. Twenty years later, Earth Day 1990 was still concerned with the needs and wants and goals of humans, but on a much larger scale—dealing with the problems of the "greenhouse effect," desertification (which reduces land available for growing crops), and massive destruction of the oxygen-producing and carbon dioxide-absorbing rain forests. The increasing rate of extinction of species seemed only secondary.

The immediate response to the "environmental crisis" included such things as recycling on a much larger scale, more attention to energy efficient building and other technologies and, on the human consciousness level, the whole Gaia movement. Although laudatory in its inception as the hypothesis by Lovelock, it very quickly became another level of abstraction: the whole earth as goddess. The underlying premises, however, were still the same. Namely, an anthropocentric emphasis on human needs—what Arne Naess labels "shallow ecology."

Perhaps it is time to turn the whole process around. Instead of us deciding what's to be done and which species will be allowed to survive, let's go to the roots of the problem. First of all, humans need to find out where they fit into the on-going process of life. If we can grasp how the interconnecting lives of different entities lead to flourishing life for the whole, then we have a start. We don't need more abstract data about the "environmental crisis" and ever more powerful technologies to fix the problems, but a sense of the living reality, the life-world in which we live.

Consider, for example, the fascinating role of mychorriza or root fungus. Without root fungus, trees cannot grow; yet the industrial growth society has grand schemes of growing vast forests of trees on farms, essentially using the methods of modern agriculture (fertilizers, pesticides, mono-species planting). But these plans never consider the "below powers" under the ground, those that actually cause the tree to flourish. It turns out that not only do trees need root fungus, but this same root fungus provides humans with the mystic experience of unity that leads to religious experience in many of the world's cultures.

Up until a few years ago the only readily accessible piece of information concerning this vast interconnection was the fact that the gourmet delight, truffles, grew on the roots of oak trees. Even here we have total dependence on the nonhuman because humans could not find these truffles and had to depend on pigs to root them out. Mushrooms, another gastronomic delight, were thought to be spontaneous growths of something which had little real use for humans and were therefore unimportant. Gradually, however, research has shown that on the contrary these lowly beings fed the gigantic trees of entire forests. These observations of the interdependent association of plants, soil and microorganisms have such revolutionary implications that they are sending shockwaves through forestry and agriculture.

Mychorriza literally means fungus root. A symbiotic association is formed between feeder roots of plants and certain fungi. According to J. L. Harley, because nutrients are exchanged in both directions within the symbiotic system of plant and fungus, neither can be considered as host. For example, many trees, such as pine and eucalyptus, must be infected by mycorrhiza fungi to survive. So close is the relationship that one research worker defined the tree as a photosynthetic appendage of mycorrhizal fungi. The tree gives the fungi some of its food produced by sunlight; the fungi provides the tree much of what it needs. For example, root fungus "assists uptake of phosphate from soils far below the minimum agricultural standard of fertility." Instead of growing trees through the vast human effort of mining, transporting by truck for hundreds of miles, and spreading by other oil guzzling machines over the fields, this tiny little root fungus does the whole thing. Still another benefit of root fungus is that it has been found to produce antibiotics, active against certain plant diseases. (Harley, the leading expert in this field, also contends that mycorrhizal fungi are important not only for reforestation but also in the establishment of productive grassland.)

These very same mycorrhizal fungi which do so much for the giant forests also produce mushrooms as their fruiting bodies, such as the famous Amanita and Boletus. Amanita is the Soma of the ancient Vedic hymns of India; thus the source of much of the mystic religions of India—Amanita mushrooms—was also the foundation for the original religion, that of the Paleolithic times, based on shamanic type religious experiences. Boletus is also a gourmet mushroom, which cost $87 a pound in French country markets a few years ago. In my own mountains they grow under spruce trees. I had noted that fact in gathering them, but had not realized until recently that, of course, they grow under spruce: they are the root fungus for spruce.

I learned more about root fungus when I inserted some in the dirt around a tiny aspen seedling which had volunteered in my yard. In Silverton people continually try to move aspen from the nearby hills into town and most often they die or linger for years in a sickly condition. Our town is 9300 feet high and there has been mining here since 1880, so there is little soil. I became a root fungus enthusiast when the root fungus allowed my little seedling to grow to the same trunk thickness as the twenty-five year old aspen trees of the nearby forest in only six years.

The lesson here is that human beings cannot do what the tiny, underground mycorrhiza do for the tree with the methods of industrialized agriculture. (We can, for example, provide phosphate fertilizers, but *only* at the cost of vast expenditures of oil with the resulting destruction of distant places and increase in climate warming.) Primitives called these unseen ones the "below powers" and respected them. In fact, the Australian aborigines do their "increase ceremonies" for all the beings of their place including fungus and biting insects because they are a part of the life of that place.

From the "below ground" powers to the "above powers"

I want to talk about the relationship between positive and negative ions and human beings. Humans pride themselves on their particular consciousness, unique to them. And yet, in reality, this consciousness is controlled by the fluctuations of the environment to the extent that differing combinations of the ions in the molecules of the air can either lead to refreshing exuberance or near suicidal depression. Who is in charge here? Humans or the invisible ions of the air?

Most of the early research in this field was done in Russia. Only in the last two decades have Americans begun to recognize the effects. When I lived in Switzerland I was delighted to find that people recognized the effects of the "bad" positive ions. Their foehn wind blows up from the south, bringing with it warm air and positive ions which cause restlessness among animals and humans, as well as headaches, bad temper, depression and inability to work. In this country there are similar winds, such as Chinooks throughout the West and the Santa Ana in California. But we are not part of a culture where continuing folk-wisdom has provided the words for such effects on humans. Positive ions are prevalent not only in bad winds but in polluted city air. Negative or "good" ions are preponderant in forests, by waterfalls and on ocean beaches. Positive and negative ions are the result of atmospheric disturbances far bigger than any human. It's nonsense to say that we are *one* with such vast beings; yet we are deeply influenced by them. Nor can we say we are part of these enormous atmospheric disturbances as they often cover thousands of miles. The reality in daily life is that we are influenced by them in the particular place where we are living. This is sometimes referred to as the bioregional "self" of which we are a part; however, I see this as part of the "substance trap" inherent in western culture. (See my *Sacred Land, Sacred Sex* for an extended discussion of the "substance trap.")

Arne Naess, the founder of Deep Ecology was deeply influenced by Spinoza's work; thus one of his central tenets is self-realization, calling for the realization of as expansive a sense of self as possible—becoming identified with the rivers, the mountains and the trees so that we care for it all as our own self. But this is still "a self" and therefore a "substance." But consider that each "self" is engaged in an on-going process called "life."

Life does not begin with our own small self and extend to the greater Self, but rather the on-going life of the whole is going on all the time. It's our role to find a niche in the process which does not detract from the fullness of life but rather adds to it even in a small way. Arne Naess once said that what he really means by the larger self-realization within the whole of nature might be called "symbiosis."

Living in place

I WANT TO give you two accounts of how humans can work within the whole of nature in their place without the immense destruction of modern technology.

The first concerns the primitive Nootka method of cutting the cedar wood used to make the planks for their houses. Those of us who use the primitives as a model are often accused of wanting to go back and live in caves with no fire. This is precisely why I choose the Nootka technique as an example. The Nootka tribe of the Northwest were true primitives, but they lived in well-built houses. To build these houses they did not engage in the vast destructive industry of modern forestry; instead, a single man, as he wandered through the forest, might see the proper kind of tree—one with no low limbs so that there were no knots and a straight lower trunk. He would first cut a narrow notch with chisels, half-way across the trunk, just above the level where the root-buttresses begin to flow out. Then he threw a line over the lowest branch and climbed up, about twenty or thirty feet above the notch. Below the lowest branch he would cut a wider notch also halfway through the trunk. Next he drove wedges downward into the trunk at the bottom of the upper notch until the wood began to split. Then he thrust a pole in, below the wedges, climbed down and went home. The weight of the pole and the action of the wind rocking the tree gradually extended the split to the lower notch, at which point the half-section of the trunk between the notches fell off and it could be used for boat or house building—but the tree was still living. This is real sustainable forestry.

The other example is a more recent illustration of "letting nature do the work." Jaquetta Hawkes gives an account of how people quarried the Stonesfield limestone for tiles in England, cutting out the thin bed of limestone by means of shafts, sometime between Michaelmas and Christmas. Then it was put in clamps and left there. "Extraordinary as it may seem, it could not be artificially split." When the frost came, usually in January, "every man in the village rallied to spread out the slabs of pendle; if it fell suddenly during the night the church bells were rung to summon the villagers." They took apart the clamps and spread out the neatly cut slabs on the grass. That's all they had to do; the frost had done the rest. If they did not get a frost that year, they had to bury the pendle deep in the soil "for if once the quarry water was allowed to escape the slabs became 'bound' and could never be split." This method of splitting limestone into tiles was done right up to the time of World War II.

In contrast with these two accounts of humans working harmoniously with nature through an abiding knowledge of local place, consider an example from my own state of Colorado: a story not of letting nature do the work but of the utter destruction of vast forests just to smelt ore to make some rich man, far away, even richer. Before the railroads made coal easily accessible much of the smelting for mines was done with charcoal. For example, near Leadville "burning pits 40 x 12 x 15 feet were excavated in the ground and spruce logs, cut to four foot lengths, were stacked in them. They were set afire and when burning sufficiently were covered with earth and allowed to smolder for about two weeks. Then the charcoal was recovered." This amount of charcoal served for only one firing in a smelter, and there were sixteen smelters operating. "Between timber for the mines, the charcoal industry and domestic firewood, it is little wonder the nearby forest was depleted." The facts concerning charcoal-making noted above come from the Leadville, Colorado area. But it was repeated all over the mining region and accounts for the way the forests look throughout the entire Front Range in Colorado. There are trees but widely scattered, stunted still and with little undergrowth. High altitude takes more than a mere 100 years to recover from such abuse.

In the arid climate of Nevada it was even worse. They used pinyon to make charcoal. It is estimated that between 400,000 and 525,000 acres of pinyon woodland were decimated for charcoal burning alone. Adding to that the wood used for cooking, constructing and mine use "would probably raise the total area to three quarters of a million acres, about an eighth of Nevada's woodland area." Nevada was not always such a treeless land! Modern humans desertified it, thereby getting rid of the Indians by depriving them of one of their prime sources of food—pinyon.

Essentially we're talking about habitat here—*a place to live.* As Eddie Nickens points out: "The purest rain won't grow crabgrass on asphalt. The cleanest air won't sustain wildlife that has nowhere to live"—or humans either. For life to continue in any particular place we must learn to pay total attention to the needs of the land itself—or rather, the whole of the on-going processes contributing to life on that piece of land. This is what primitive hunters did. As Paul Shepard explains: "Since they know nature well enough to appreciate how little they know of its enormous complexity, hunter-gatherers are engaged in a vast play of adventitious risk. . . . Their myths are rich in the strangeness of life, its unexpected boons and encounters, its unanticipated penalties and mysterious rewards. . . . Their lives are committed to the understanding of a vast semiosis, presented to them on every hand, in which they are not only readers but participants."

The ancient Chinese called this process *li*, often translated as pattern. The earliest meaning of *li* came from the pattern in which fields were laid out for cultivation in order to conform to the typography of the land; hence the earth was considered the ordering principle of a particular place. Later, the word *li* came to refer to patterns in things—the markings in jade or the grain in wood. The Neo-Confucianist Chu Hsi used the word *li* for the principle of cosmic organization which ran through all levels of Being. The Chinese did not feel that either God or law was the ordering principle; rather they saw the universe as a hierarchy of parts and wholes working in harmony—each according to its own *li*, or pattern, but all fitting together in the great pattern. Within each place on earth a particular pattern develops from the interactions of sun, gravity, and space—the framework into which each fits. Humans, too, can begin to fit into the pattern of the place but not by conscious willing; instead, living in place begins with paying attention to all the natural entities around us, and then weaving myth and story out of that lore. The place gives us the *li*, the pattern; "the story" is a narrative account of how we fit into that pattern.

Tribal people, in general, as well as the pre-Socratics and the early Chinese all make use of patterns in their basic thinking. We tend to call tribal people illiterate, but that is because we do not realize that their patterns, songs and dances are valid literature and accurate recording systems.

Gregory Bateson, who has been called the seminal thinker of this century, has this to say about pattern or *li*. "We have been trained to think of patterns, with the exception of those of music, as fixed affairs. It is easier and lazier that way but, of course, all nonsense. In truth, the right way to begin to think about the pattern which connects is to think of it as *primarily* (whatever that means) a dance of interacting parts and only secondarily pegged down by various sorts of physical limits and by those limits which organisms characteristically impose."

Gary Snyder, our foremost poet in Deep Ecology, also tells us of the importance of pattern:

> So there is work to be done in the matter of knowing where we are, the old American quest, which I share with all of you, for an identity, a sense of place. To know the place well means, first and foremost, I think, to know plants, and it means developing a sensitivity, an openness, an awareness of all kinds of weather patterns and patterns in nature.

This "awareness" will teach us to understand our proper niche as humans within the on-going life of our particular place on earth. Then, we, too, can *know*—as tribal people do:

> *The ear for the story and the eye for the pattern were theirs; the feeling was theirs; we came out of this land and we are hers.* —Leslie Silko, Laguna Pueblo

*
**

Weak Anthropocentric Intrinsic Value

Eugene Hargrove

Introduction

PROFESSIONAL ENVIRONMENTAL ethics arose directly out of the interest in the environment created by the first Earth Day in 1970. At that time many environmentalists, primarily because they had read Aldo Leopold's essay, "The Land Ethic," were convinced that the foundations of environmental problems were philosophical.[1] Moreover, these environmentalists were dissatisfied with the instrumental arguments based on human use and benefit, which they felt compelled to invoke in defense of nature, because they thought these arguments were part of the problem. Wanting to counter instrumental arguments in some way with non- or even anti-instrumental arguments, and unable to think of anything else to say, they began wistfully suggesting that perhaps nature had or ought to have rights.[2] When professional environmental ethics came into its own in the early 1980s, rights for nature were one of the first subjects to be debated in detail.[3] Unfortunately, however, no one could come up with a theory to support such rights attributions. Nevertheless, because rights had been invoked by environmentalists to challenge the preeminent role of instrumental value arguments, and because the field of environmental ethics developed in support for environmental concerns and arguments, environmental ethicists turned to an examination of noninstrumental or intrinsic value arguments for the preservation of nature.

As these investigations progressed, it soon became clear that most environmental ethicists, and indeed most environmentalists, did not believe that traditional intrinsic value, for example, the kind of intrinsic value which is attributed to art, was an adequate counter to instrumental value.[4] To find a kind of intrinsic value that could trump instrumental value, in the way that rights can, they started looking for nonanthropocentric intrinsic value. This search, unfortunately, has been a confusing one because of definitional problems with the word *nonanthropocentric*. A nonanthropocentric value was simply assumed to be the opposite of an instrumental value, making *anthropocentric* for all practical purposes a synonym for the word *instrumental*.[5] In environmental policy, there is perhaps some basis for such a definition, since nearly all arguments, economic and otherwise, are formulated routinely in terms of instrumental value to human beings. Nevertheless, *anthropocentric* is not and has never been a synonym for *instrumental*. It simply means "human-centered," and refers to a human-oriented perspective, seen from the standpoint of a human being.[6] This confusion results from an assault on intrinsic value undertaken by pragmatists at the beginning of this century, who tried to eliminate intrinsic value talk and substitute instrumental value talk across the board. Although no one today can remember, let alone clearly formulate, the pragmatic reasons for abandoning intrinsic value, the idea that intrinsic value is an unnecessary concept has managed to trickle-down to the level of ordinary people, who now *believe* that only instrumental value arguments work but nevertheless *wish* that it were not so.[7] Moreover, the situation has been further aggravated by another even more widely believed corollary of logical positivism which has also successfully trickled-down, viz., the belief that all value judgments are meaningless, arbitrary, subjective, irrational expressions of emotion.[8]

There are actually two kinds of nonanthropocentric intrinsic value theory, an objectivist version, of which Paul Taylor and Holmes Rolston, III are the most prominent proponents, and a subjectivist version proposed by J. Baird Callicott. In this essay, I discuss both objective and subjective nonanthropocentric intrinsic value in contrast to a counter position, weak anthropocentric intrinsic value. I argue (1) that objectivist nonanthropocentric intrinsic value theory requires and is complemented by weak anthropocentric intrinsic value theory, (2) that the most plausible version of subjectivist nonanthropocentric intrinsic value theory is actually a form of weak anthropocentric intrinsic value theory, (3) that a weak anthropocentric intrinsic value theory is superior to a weak anthropocentric value theory based on pragmatic instrumentalism, and (4) that most nonanthropocentric value theories are in various ways really anthropocentric.

Objectivist nonanthropocentric intrinsic value theory

ANY EXAMINATION OF objective nonanthropocentric value should begin with an examination of its practical role in environmental ethics and indeed ethics in general. In the three periods of Western civilization, there have been two very different approaches to ethical decision making. The first is a virtue approach, according to which people are trained to develop good moral character on the assumption that moral persons will act morally. The second is a rule approach which aims at establishing a set of universal rules that are to be followed without deviation. The former is the characteristic approach of the ancient and medieval periods, the latter of the modern period. As I have argued elsewhere, the effect, if not the intention, of the rule approach is to limit the range of ethical decision making, turning moral decisions into either/or situations, in which following a rule, without regard to consequences, is what being ethical is all about.[10] This approach has gained support in the modern period because of "fear that the open form in which decisions naturally and normally take place will allow unscrupulous or weak moral agents to waver and modify principles to their own immoral advantage."[11] Efforts to establish unmodifiable universal rules for ethical decision making usually come at the end of a period of emotivism, *Sturm und Drang* followed by Kantianism in the last century, and positivistic emotivism followed by prescriptivism in this century.[12]

The motivation behind the quest for an objective nonanthropocentric value theory seems to be of this kind, for objective intrinsic value is supposed to be independent of and override individual human judgment and the relative and evolving cultural ideals, which though currently supportive of nature preservation, might change leaving nature without moral defense.

This concern is explicit in the writings of Paul Taylor, for example, who, citing Mark Sagoff's anthropocentric symbolic value theory, argues that he wants to develop a nonanthropocentric value theory because anthropocentric value is "entirely relative to culture: If a particular society did not hold ideals that could be symbolized in nature and wildlife (for example, if it happened to value plastic trees more than real ones), then . . . there would be no reason for that society to preserve nature or protect wildlife."[13] On analogy with the distinction between the basic rules of a game and rules of good play, I have distinguished between two kinds of rules, constitutive rules and nonconsititutive rules.[14] The first are rules which must be followed without deviation. Not doing so automatically produces an immoral act. The second are guides which may or may not be followed depending on specific circumstances in specific situations. Objective nonanthropocentric intrinsic value is supposed to play a limiting role similar to that of constitutive rules. Recognition of the existence of these values in nature, independent of human judgment and culturally evolved values, putatively automatically triggers specific moral behavior.

Since the field of environmental ethics first began, environmental ethicists have always been very much aware that educated attitudes toward nature in the Middle Ages, according to which nature is not beautiful, would not have been supportive of twentieth-century-style environmental concern, and that it is possible that current supportive values might change again.[15] This concern is also shared by most environmentalists. When I speak on behalf of weak anthropocentric value, I am frequently asked by environmentalists, "But what if people change their minds and stop thinking that nature is beautiful?" The possibility that our culturally evolved environmental values might change is, of course, a serious concern, given that these values, though derived from aesthetic tastes in art, primarily, from landscape painting, but also from nature poetry, landscape gardening, and natural history science, have now been out of fashion in art circles for more than a century. Nevertheless, although constitutive values might produce a trumping effect similar to rights for nature, if and when people ever come to be convinced that such values exist, I doubt that attempts to persuade ordinary people that such values exist independently in nature is a wise long-term environmentalist strategy. The best way, in my view, to deal with this concern is actively to defend these values as part of our cultural heritage, not to try to develop a metaphysical/epistemological theory of objective nonanthropocentric intrinsic values that constitutively trumps individual judgment and culturally evolved values. Defending these values as culturally derived values will focus attention on the merits of these values and presumably help strengthen them. In contrast, an objective nonanthropocentric value theory will draw attention away from their merits as cultural values (the point of the objectivist effort) and refocus it on metaphysical and epistemological issues that ordinary people are unlikely to understand or be persuaded by. While this approach might succeed for a time in freezing current values (by embedding them in our constitutive moral framework), it certainly cannot strengthen them (since they are supposed to be independent of human judgment and culture, humans should not be able to alter them one way or the other). In the event that aesthetic tastes toward nature do start to change (for example, "catching up" with those of modern art), a spirited defense of these values as cultural values might prevail. Whether an argument that these values are independent of and trump human judgment, and therefore cannot be changed, would succeed, however, seems less likely, given that ordinary people can easily refute the argument without providing any counterargument of their own simply by changing their values.

To be sure, there is a lot of room for legitimate nonanthropocentric value theory in environmental ethics, for most values, whether instrumental or intrinsic, *are* independent of human judgment. At the most general level, four kinds of values are possible: (1) nonanthropocentric instrumental value, (2) anthropocentric instrumental value, (3) nonanthropocentric intrinsic value, and (4) anthropocentric intrinsic value. In environmental matters nonanthropocentric instrumental values, concerning the instrumental relationships of benefit and harm between nonhuman plants and animals, are quite common and completely uncontroversial. Such values, which can easily be converted into facts, are indeed discovered in the world and are independent of human judgment. One thing in nature either instrumentally benefits other things or it does not, regardless of what humans think about it and whether or not humans even know that these instrumental relationships exist. Bamboo is either instrumentally valuable to pandas or it is not. What we believe, know, and how we value it makes no difference. Anthropocentric instrumental value judgments, if they are simply the same relationships applied to humans, are likewise common and uncontroversial. Fluoride is either instrumentally valuable to humans or not, whether we humans know it, believe it, or value it. Even if *anthropocentric* is taken to require human judgment, producing judgments or beliefs that one thing is instrumentally valuable or harmful to another thing (human or otherwise) without (adequate or complete) factual evidence, such anthropocentric instrumental value judgments are also uncontroversial when they can be converted into the form of scientific hypotheses, making them potential or possible facts, and irrelevant (rather than controversial) when they are merely whimsical or irrational expressions of belief. Smoking tobacco is widely judged to be instrumentally harmful to human health, even though it is sometimes scientifically disputed by tobacco companies; in contrast, even though chicken broth is generally judged to be instrumentally beneficial to humans with colds, this judgment is not supported by medical evidence. Similarly, the basic nonanthropocentric maneuver, notably exemplified by Paul Taylor and Holmes Rolston, III, defining living organisms as centers of purpose in accordance with Aristotle's ends/means distinction, is also noncontroversial, if all that is claimed is that these entities have sakes or goods of their own (independent of human interests) and that they are using nature instrumentally for the benefit of their own sakes, which are then defined as (intrinsically valuable) ends (to them).[16] Because such intrinsic value assignments are also a matter of discovery rather than judgment, they too can be treated as disguised facts; we can discover facts through scientific research that such and such kinds of organisms do or do not instrumentally use other parts of nature for their own ends. It is actually only anthropocentric intrinsic value assignments, judgments made by humans that such and such living and nonliving entities are noninstrumentally (intrinsically) valuable, that fully and truly depend upon human judgment, rather than mere discovery, and are not ever reducible to facts or scientific hypotheses.

There seem to be two basic reasons why objectivist nonanthropocentric value theorists object to doing away with or at least radically deemphasizing anthropocentric intrinsic value judgments. First, as noted above, they want values that operate much like constitutive rules in order to trump anthropocentric instrumental values. Second, they hold that there can only be (or should only be) one kind of intrinsic value.[17] It is this second reason which I find most problematic. Given that there are many kinds of instrumental value, nearly all of which have something to do with environmental ethics, it seems strange to me that anyone would want to claim that there can only be one kind of intrinsic value, or, if it is acknowledged that there may be more than one kind, that only one kind is relevant to environmental ethics.[18] It is almost as if there is a competition between various conceptions of intrinsic value such that recognition of one kind of intrinsic value, anthropocentric intrinsic value, somehow damages the other, nonanthropocentric intrinsic value. In opposition to this strange view, I want to argue here that anthropocentric intrinsic value judgments, rather than being in competition with nonanthropocentric intrinsic values, are absolutely essential if humans are to muster any environmental concern about nonhuman living centers of purpose (as well as many other natural entities) objectively existing out in the world.

Paul Taylor, a proponent of nonanthropocentric intrinsic value, which he calls *inherent worth*, has made a noteworthy attempt to distinguish and separate nonanthropocentric and anthropocentric intrinsic value. After briefly mentioning instrumental value, commercial value, and merit or excellence, Taylor offers three kinds of intrinsic value: the immediately good, the intrinsically valued, and inherent worth. After defining (1) *instrumental value*, (2) *commercial value*, and (3) *merit or excellence*, none of which are important for the purposes of this paper, he defines (4) *the immediately good* as "any experience or activity of a conscious being which it finds to be enjoyable, satisfying, pleasant, or worthwhile in itself," noting that it "is sometimes called *intrinsic value*." He then provides two long definitions for (5) *the intrinsically valued* and (6) *inherent worth*, complete with examples, which I quote here in full:

(5) *The intrinsically valued.* An entity is intrinsically valued in this sense only in relation to its being valued in a certain way by some human valuer. The entity may be a person, animal, or plant, a physical object, a place, or even a social practice. Any such entity is intrinsically valued insofar as some person cherishes it, holds it dear or precious, loves, admires, or appreciates it for what it is in itself, or so places intrinsic value on its existence. This value is independent of whatever instrumental or commercial value it might have. When something is intrinsically valued by someone, it is deemed by that person to be worthy of being preserved and protected because it is the particular thing that it is. Thus, the people of a society may place intrinsic value on a ceremonial occasion (the coronation of a king), on historically significant objects (the original Declaration of Independence) and places (the battlefield at Gettysburg), on ruins of ancient cultures (Stonehenge), on natural wonders (the Grand Canyon), and of course on works of art. Intrinsic value may also be placed on living things, which then are intrinsically valuable to (have intrinsic value for) the human valuers. A pet dog or cat, an endangered population of rare plants, or a whole wilderness area can be considered worth preserving for just what they are. Finally, anyone we love and care about has this kind of value for us. From a moral point of view, correlative with intrinsically valuing something is the recognition of a negative duty not to destroy, harm, damage, vandalize, or misuse the thing and a positive duty to protect it from being destroyed, harmed, damaged, vandalized, or misused by others.

(6) *Inherent worth.* This is the value something has simply in virtue of the fact that it has a good of its own. To say that an entity has inherent worth is to say that its good (welfare, well-being) is deserving of the concern and consideration of

all moral agents and that the realization of its
good is something to be promoted or protected as
an end in itself and for the sake of the being
whose good it is. Since it is only with reference
to living things (humans, animals, or plants) that
it makes sense to speak of promoting or protecting their well-being and of doing this for their
sake, the class of entities having inherent worth
is extensionally equivalent to the class of living
beings.[19]

The immediately good, as Taylor defines it, is the product of an
instrumental relationship between a "good" object and a human
being exposed to it. As Taylor presents it here (rightly, I think),
it does not involve human judgment, merely spontaneous emotional reaction to sensory events triggered by an external source.
Taylor's discussion of the intrinsically valued is a good presentation of what I mean by anthropocentric intrinsic value, intrinsic
value assigned or attributed by a human being or a group of
human beings from an independent ahistorical human perspective
or from a culturally dependent historical human perspective.
Taylor's inherent worth is a fairly standard version of
nonanthropocentric intrinsic value attributed to entities which
have a sake, instrumentally use other parts of nature for their
own intrinsic ends, or, as Taylor puts it, have goods of their own.
    From my perspective, the key issue concerning Taylor's
account of intrinsic value is whether the intrinsically valued and
inherent worth can be radically separated. If they can't, then a
creature having inherent worth or a good of its own is simply one
of the various kinds of things that humans may, if they wish,
collectively or individually, value intrinsically. Immediately
following the definitions and discussions of the various kinds of
values listed, Taylor writes, "When a living thing is regarded as
possessing inherent worth, it is seen to be the appropriate object
of the moral attitude of respect. This kind of respect, 'recognition respect,' should not be confused with the attitudes of love,
admiration, and appreciation directed toward entities that are
intrinsically valued in sense (5) above."[20]
    Two questions occur to me about these two sentences and
the discussion that follows them. First, has Taylor shown that

respecting something is different than intrinsically valuing it? At a minimum, Taylor has correctly pointed out that respect is different from love, admiration, and appreciation. One can respect one's enemies, for example, without loving them. He has, however, as far as I can determine, merely asserted, but not shown, that respect is something so different that it should not be placed together with love, admiration, and appreciation as a form of intrinsic valuing. Second, when "it is seen" (passive voice) that such and such entity is "the appropriate object of respect," why doesn't it simply mean that some human decides to value such and such an entity intrinsically, that is, by means of an act of judgment, attributes (active voice) intrinsic value to the thing in accordance with some personal or culturally derived standard? The use of the passive voice together with the introduction of the term *recognition respect* seems to suggest that there is no human value judgment involved, that the human simply "sees" or discovers that the entity has a good of its own, automatically triggering a feeling of respect analogous to the instrumental relationship characteristic of the immediately good.[21] Such an account of respect as automatic recognition bypassing human judgment seems to me to be implausible.

Consider the alien monsters of the films *Alien* and *Aliens*, which require the deaths of many other living creatures, indifferently including humans, in order to reproduce and survive as a species. The newly hatched alien monster emerges from its (his or her) egg and immediately enters a host organism, which it keeps alive and feeds on while continuing its development. When it no longer needs the host, it explodes out of the chest of the organism, killing the host with some fanfare. Up until this moment, the human host is conscious and aware of his or her situation. Because these creatures have goods of their own, according to Taylor, they have inherent worth. But does it follow from the fact that such a creature has a good of its own that we humans are *required*, in accordance with Taylor's definition of *inherent worth*, "to say that its good (welfare, well-being) is deserving of the concern and consideration of all moral agents and that the realization of its good is something to be promoted or protected as an end in itself for the sake of the being whose good it is"? I think not. Rather, the (nonanthropocentric) inherent worth or good of its own of the alien monster produces a concern and consideration if and only if a human decides (or humans collectively decide) to intrinsically value the creature, thereby producing, in accordance with Taylor's definition of *the intrinsically valued*, "the recognition of a negative duty not to destroy, harm, damage, vandalize, or misuse the thing and a positive duty to protect it from being destroyed, harmed, damaged, vandalized, or misused by others." The circumstances under which the necessary anthropocentric intrinsic value attribution might be generated are fairly limited. First, the human and all other humans (and all other inappropriate potential hosts) would have to be safe from the organism (a point with which, I believe, Taylor would agree). Second, the creature would have to be in its natural ecosystem, rather than in some other ecosystem, where it would be regarded as an improper (and very destructive) pest. In the two films, two characters (an android and a human) do value alien monsters and try to preserve them, but they do so because of their instrumental value as military weapons (such creatures once released could presumably destroy all human life fairly easily, given the speed with which they are able to reproduce). It is true that humans might quickly develop a (healthy) respect for such creatures, but this respect would not be based on the recognition that these creatures have goods of their own, but rather would be out of fear, having recognized that the creatures are very dangerous.

Note that I am not claiming that a creature's good of its own is irrelevant to the moral concern of humans, merely that the fact that a particular creature has a good of its own is not enough automatically to produce moral behavior on behalf of the creature. After discovering that something has a good of its own, the human or humans must decide to intrinsically value it, and in doing so, the specter of cultural (and historical) relativity reappears, which was avoided by omitting mention of the need for intrinsic value attribution.

Note also that even though objectivist nonanthropocentrists are committed to doing away with anthropocentric intrinsic value, as an anthropocentrist, I have no similar desire to bring the quest for nonanthropocentric intrinsic value to an end. I hold that "weak anthropocentrism" (the view, as I define it, that *anthropocentric* does not simply mean instrumental) can serve environmental ethics well until such time, if ever, that a convincing nonanthropocentric theory appears that will sweep strong and weak anthropocentrism aside.[22] Moreover, I hold that whether the quest succeeds or fails, it will further enhance anthropocentric intrinsic value theory by providing new grounds for intrinsically valuing nonhuman life anthropocentrically. I confess, however, that I am not very optimistic that a nonanthropocentric theory will be successfully formulated because the search for nonanthropocentric intrinsic value seems to me to be comparable to a Kantian search for actual objects in the noumenal world. To succeed the anthropocentrists apparently need to go beyond valuing based on the human perspective, which seems impossible.

Note, finally, that even if a persuasive nonanthropocentric theory of intrinsic value can be constructed, environmentalists and environmental policy and decision makers will probably still need to make some anthropocentric intrinsic value judgments. Currently, nonanthropocentric theory, based as it is on the goods of individual organisms, leaves nonliving natural objects out of the moral account. As Taylor notes in his definition and discussion of inherent worth, cited above, the "class of entities having inherent worth is extensionally equivalent to the class of living beings." Thus, nonliving objects can only be defended on the grounds that they are instrumentally valuable to living centers of purpose that use them for their own intrinsically valuable ends.

Such an approach is, however, woefully inadequate with regard to the kinds of objects that I was principally interested in protecting when I undertook work in the field of environmental ethics—caves. Strictly speaking, caves are not objects at all. Paraphrasing Wittgenstein on pain, they are not a something but not a nothing either, that is, they are hollow spaces in layers of sediments. As a cave conservationist, I quickly learned that it is difficult to produce winning arguments to protect caves in terms of the creatures who live in them. Bats, worms, insects, and blind fish, though less distasteful than alien monsters, generate little preservationist concern or sympathy. Many people, for example, foolishly think that it would be a good idea for all bats to be killed on sight. The strongest arguments for protecting caves, in contrast, depend upon the willingness of humans to act so as to preserve natural beauty, which in turn depends upon intrinsic valuing on the model of the aesthetic appreciation of art objects. As long as nonanthropocentric theorists concentrate on the class of living objects, nonliving beauty will continue to be left out, and will require anthropocentric intrinsic value attributions in order to receive any protection at all. Although protection of living organisms may require the partial abandonment of an anthropocentric perspective (to the degree that that may be possible), protection of nonliving objects requires a return to an anthropocentric perspective, unless these objects are to be valued instrumentally only. That approach, however, would run counter to our basic environmental intuitions and unnecessarily abandon very strong arguments for nature preservation, given that the historical/cultural foundations of environmental ethics (going back three hundred years) are primarily aesthetic.[23]

Because I consider my conception of intrinsic value to be (1) a necessary element in the valuing of nonanthropocentric intrinsic value and (2) complementary to nonanthropocentric valuing, since only anthropocentric intrinsic value can be applied to nonliving objects, I have no quarrel with the objectivist nonanthropocentrists and their quest as long as they do not insist that (1) only living entities with goods of their own matter morally and that (2) they matter in a way that is independent of human (anthropocentric) intrinsic valuing.

Accepting anthropocentric and nonanthropocentric value as complementary and interrelated, moreover, can even improve some objectivist nonanthropocentric theories. For example, Rolston's narrow focus on an organism's good ultimately denies a central role to natural beauty in his theory because it is too subjective to fit in comfortably with his desire to develop an objective approach. When Rolston finally reaches natural beauty in his book, he is only prepared to go through the motions of a discussion, and ultimately has little to say about it, because he can't effectively tie it into his objective theory: "With beauty we cross a threshold into a realm of higher value; the experience of beauty is something humans bring into the world."[24] According to Rolston, the value focus in natural systems is on the organism with its good of its own.[25] In this context, he divides the world up into beholders of value (humans) and holders of value (organisms with goods of their own), the value that the beholders behold. To find a way to include value beyond holders and beholders, Rolston adds value producers (the ecosystem) which provide (instrumental) support for organisms, the value holders. According to Rolston, the value produceded by ecosystems, *systemic value*, flashes back and forth between instrumental and intrinsic value, on the model of particles and waves of light.[26] In addition, however, to maintain the dualism between holders and beholders Rolston goes on to maintain that an ecosystem as a value producer is really a value holder "in the sense that it projects, conserves, elaborates value holders (organisms)."[27] When he finally turns to beauty directly, he addresses two fairly trivial questions, whether nothing in nature is beautiful and whether everything in nature is beautiful, and argues that the answer is somewhere in between, if we accept that the sublime is also beautiful.[28] The insignificant role of natural beauty in his theory, brought about by the quest for nonanthropocentric intrinsic value, is truly unfortunate, because, as is evident in virtually every sentence Rolston writes, no philosopher has a better feel for and appreciation of natural beauty than he does. To provide a place in his theory for his own aesthetic values, all that Rolston needs to do is simply to go beyond nonanthropocentric intrinsic value, defined in terms of the good of its own of an organism, and reintroduce anthropocentric intrinsic valuing into his theory. (This is advice that I also recommend to all other nonanthropocentrists who have pointlessly lost touch with natural beauty by trying to be too objective.)

Subjectivist nonanthropocentric intrinsic value

ALTHOUGH THERE IS only one subjective nonanthropocentric value theorist, J. Baird Callicott, there are two theories. First, Callicott has argued that intrinsic value is conferred on the natural world by humans valuing it for itself. Second, he has flirted with the Self-realization approach of deep ecology, in which nature comes to have intrinsic value because a Self-realizing human becomes one with it. I deal with the Self-realization theory only in passing and focus my discussion on the first theory.

Callicott summarizes his Self-realization approach as follows:

> Now *if* we assume, (a) . . . that nature is one
> and continuous with self, and (b) that self-inter-
> ested behavior has a prima facie claim to be at the
> same time rational behavior, then the central
> axiological problem of environmental ethics, the
> problem of intrinsic value in nature[,] may be
> directly and simply solved. If quantum theory
> and ecology both imply in structurally similar
> ways, in both the physical and organic domains
> of nature, the continuity of self and nature, and *if*
> the self is intrinsically valuable, then nature is
> intrinsically [valuable]. If it is rational for me to
> act in my own interest, and I and nature are one,
> then it is rational for me to act in the best interest
> of nature.[29]

This approach is said to produce nonanthropocentric value,
rather than anthropocentric value, because the human self is
only a small part of nature as a whole, which is the Self with a
capital *S*. Nevertheless, an alternative interpretation is available, according to which self-realization is anthropocentric and
nothing more than Cartesianism commandeered for environmental purposes. Note that nature acquires (or borrows) its
intrinsic value from the human self, which is established on
supposedly noncontroversial traditional grounds (the uncritical
belief that humans have intrinsic value).[30] Seen in this way, the
intrinsic value of nature is a product of the humanizing of
nature, and the model is nineteenth-century idealism and solip-
sism, arising out of the Cartesian puzzle about whether humans
can know that the world exists.[31] Because such humanizing is
frequently considered one of the causes of our environmental
problems, this approach is suspect.

Turning to the other theory, Callicott has also claimed that
values depend entirely on human judgment, that there are no
values in the world until they are imposed by humans: "Value is,
as it were, projected onto natural objects or events by the
subjective feelings of observers. If all consciousness were anni-
hilated at a stroke, there would be no good and evil, no beauty
and ugliness, no right and wrong; only impassive phenomena
would remain."[32] Nevertheless, Callicott insists that he can

develop a nonanthropocentric intrinsic value theory within this domain of humanly generated values, which he calls his "truncated" theory of intrinsic value:

> I concede that, from the point of view of scientific naturalism, the *source* of all value is human consciousness, but it by no means follows that the *locus* of all value is consciousness itself or a mode of consciousness like reason, pleasure, or knowledge. In other words, something may be valuable only because someone values it, but it may also be valued for itself, not for the sake of any subjective experience (pleasure, knowledge, aesthetic satisfaction, and so forth) it may afford the valuers. Value may be subjective and affective, but it is intentional, not self-referential. For example, a newborn infant is of value to its parents for its own sake as well as for the joy or any other experience it may afford them. In and of itself an infant child is as value-neutral as a stone or a hydrogen atom, considered in strict accordance with the subject-object/fact-value dichotomy of modern science. Yet we still may wish to say that a newborn infant is "intrinsically valuable" (even though its value depends, in the last analysis, on human consciousness) in order to distinguish the *noninstrumental* value it has for its parents, relatives, and the human community generally from its actual or potential instrumental value, the pleasure it gives its parents, the pride it affords its relatives, the contribution it makes to society, and so forth. In doing so, however, "intrinsic value" retains only half its traditional meaning. An intrinsically valuable thing on this reading is valuable *for* its own sake, *for* itself, but it is not valuable in itself, that is, completely independently of any consciousness, since no value can, in principle, from the point of view of classical normal science, be altogether independent of a valuing consciousness.[33]

According to Callicott, this view is nonanthropocentric, rather than anthropocentric, because human valuers (anthropogenically) value things other than themselves, and it is intrinsic, because human valuers value these other things for their own sakes. Nevertheless, it is truncated because although human valuers value things *for* themselves, nonhuman things are not valuable *in* themselves, because there is no objective nonanthropocentric intrinsic value *in* nature.

Although there are a number of problems with this position, fairly minor adjustments could transform it into a weak anthropocentric position. First, it is a bit of an overstatement to claim that *all* value depends on the subjective feelings of human observers and that value does not exist in nature unless it is projected on it by human valuing. As I indicated at the beginning of this paper, there are many kinds of nonanthropocentric instrumental values that exist in nature independent of human judgment. In addition, I see no reason why Callicott should dismiss entirely the claim that nonhuman creatures have independent intrinsic value in the sense that they have goods of their own, that is, that they are ends instrumentally using their environment for their own sake. Problems arise with this kind of value only when it is declared to be the only kind of intrinsic value or the only kind that matters. Just as objectivist nonanthropocentrists need to acknowledge that human intrinsic valuing takes place and matters, subjectivist nonanthropocentrists need to acknowledge that nonanthropocentric intrinsic value, the good of its own of an organism, even though it exists independently of human valuing, is something that humans can (whenever they so choose) value for its own sake. Finally, I question whether humans are the only beings who value nature. Given that Callicott accepts a Darwinian evolutionary approach, it seems strange that he sometimes suggests that while humans can value (like or dislike, for example) some parts of nature, nonhumans cannot.

Second, I find it questionable whether Callicott's subjectivist position should be called nonanthropocentric. If it is true that "the *source* of all values is human consciousness" and that value "is, as it were, projected onto natural objects or events by the subjective feelings of observers [that is, humans]," then all value (both instrumental and instrinsic) is centered *in* humans and radiates outward from humans *onto* things in nature. Given that I believe that other creatures also sometimes value (both instrumentally and intrinsically), Callicott's position is, in my terminology, a slightly stronger (weak) anthropocentrism than my own. For him to insist that his position is nonanthropocentric creates unnecessary and pointless confusion.

Third, I consider Callicott's position to be overly subjective. As I have argued elsewhere, human values are not entirely dependent upon the arbitrary value preferences of individuals.[34] In an Aristotelian sense, there are cultural values that are the product of social evolution. These values are not entirely subjective. At any given moment in the history of a particular society they can be objectively identified and described. Moreover, in most cases they are the foundation for the values of individual people. It is no accident that nearly all people in a particular society share the same values. They pick them up as children without formal teaching. They are the context and starting point out of which individual differences develop. Simply to call these social values subjective misrepresents their very substantial objective character.

Pragmatic instrumentalism

The most serious attack on anthropocentric intrinsic value theory, and indeed intrinsic value in general, comes from the pragmatic instrumentalists, who like the pragmatists at the turn of the century, mentioned at the beginning of this paper, want to convert such values into instrumental terms.[35] Rather curiously my chief quarrel is with my fellow weak anthropocentrist, Bryan Norton.[36] His arguments for denying that humans can (or should?) make anthropocentric intrinsic value judgments seem to be (1) that ordinary people find the word *intrinsic* confusing because it sounds mystical and (2) that presenting anthropocentric intrinsic value judgments in instrumental terms simplifies value theory, making ethics easier for ordinary people to understand.

Concerning the second claim, while I do not deny that it is certainly possible to simplify value theory in this way, most environmentalists, as noted above, have been unwilling to accept this simplification because they find instrumental presentations of noninstrumental values inappropriate and demeaning (to the natural entity), as evidenced, for example, by the fact that they still want a rights theory for nature.[37] When an aesthetic intrinsic value judgment is converted into instrumental terms, the person having the aesthetic experience is depicted as using natural scenery as a trigger for feelings of pleasure. When these feelings of pleasure are then compared with the other instrumental values that can be obtained, for example, by clear-cutting or strip-mining, the value of the aesthetic experience then appears trivial, ridiculous, and indeed indefensible.[38] In short, anthropocentric instrumental scientific and aesthetic values fail to mimic the desired trumping effect of rights theory over basic instrumental values because they are treated either as questionable or inferior to basic instrumental values.

With regard to the first claim, people are confused about intrinsic value only because they have become disoriented by the trickle-down effect of the pragmatic instrumentalist attack on intrinsic value at the beginning of this century. Before pragmatism created the confusion about the relationship of intrinsic and instrumental value, the distinction was clear and serviceable to ordinary people. The confusion caused by the blurring of this distinction is, therefore, not adequate grounds for continuing the confusion by insisting on the pragmatic instrumentalist conversion of intrinsic value into instrumental value in opposition to the clear calls by environmentalists, policy makers, and the general public for a return to intrinsic value theory. Norton has provided no evidence that we have reached a point of no return. Quite to the contrary, the dissatisfaction with instrumental arguments among environmentalists, policy makers, and the general public suggests that a return to straightforward intrinsic value talk is probably the easiest solution.

The way Norton converts intrinsic value into instrumental value is itself very confusing and, I believe, limited. According to Norton, natural objects should be valued because they have *transformative value*, value which transforms humans or changes their lives. As far as I am able to tell, this transformative value is similar to, and perhaps a general label for, what Rolston means by "character building value," "therapeutic value," and "religious value."[39] If it is, then it is too limited as a replacement for anthropocentric intrinsic value, for I do not see that the intrinsic valuing of some natural object necessarily involves an instrumental transformation of the human triggered by the object valued. Many intrinsic value attributions in art, for example, depend upon the application of historical aesthetic standards that at various times go out of fashion and, as a result, no longer move viewers emotionally (transform their lives). Although it is universally agreed that the Mona Lisa is worthy of being intrinsically valued, many, indeed, most people have no idea why it is considered to be such a valuable painting, and, lacking detailed training in the history of art, are not transformed by it when they see it. Should we say that it continues to be valuable because it once transformed people's lives and perhaps has the potential to do so again? If the value of the Mona Lisa were calculated in such terms, I think the concern of the nonanthropocentrists about the cultural relativity of anthropocentric value judgments would be justified. In real life, however, the value of a painting does not depend on the occurrence of particular emotional experiences in the general public. Rather it depends on the judgment of experts who interpret social ideals, the equivalent of the perception of Aristotle's "good man." Precise aesthetic judgments, comparable to those provided by art critics, can also be obtained by consulting professional nature interpreters, naturalists, and most environmentalists.

In practice, there is nothing confusing (or mystical or metaphysical) about anthropocentric intrinsic value judgments. Consider the value of an ornamental knife, made of precious metals and covered with jewels. Because the knife can function as a knife, it has the normal instrumental value that any other knife would have, not simply potentially, but also actually. Nevertheless, because the knife is beautiful, the judgment of the owner and others who take time to consider the matter will likely be that it is too beautiful (or good) to use (assuming that using the knife will mar its beauty). This judgment, to value the knife for its beauty, rather than its use, involves no confusing detours into metaphysics or mysticism. All that is required is an act of judgment. An individual or group of individuals or a society decides. In real life, it does not even require a defense. We do not customarily begin an instrumental argument with a general defense of instrumental arguments, for example, that instrumental value exists. We simply present the instrumental argument within the implicit framework in which such arguments are generally acceptable. The same is true of anthropocentric intrinsic value arguments. We do not need to begin such an argument with a proof that intrinsic value exists, for the existence of this kind of intrinsic value is not an issue. This kind of intrinsic value is the product of human valuing, human decision making, and everyone already knows what valuing, deciding, and judging means. The issue in such an argument is not the mental process (or the metaphysical status of the value produced), but whether the value judgment is an appropriate one, that is, in accordance with recognized social standards and ideals. The person who decides that the knife should not be used will not justify his judgment by claiming that the knife is intrinsically valuable, but rather by pointing out that the knife is beautiful. Talk about intrinsic value will occur only if someone chooses to claim that beauty is itself a use (an instrumental value), invoking the pragmatic instrumentalist maneuver. At this point, the answer could be a nonanthropocentric excursion into metaphysics; however, a simple reminder that humans are fully capable of valuing things noninstrumentally and have been doing so for thousands of years is really all that is needed.

The reintroduction of anthropocentric intrinsic value judgments is not only a simple matter, but also a very useful one. As I indicated above, the reduction of intrinsic value to instrumental terms demeans and trivializes it, giving a counterintuitive advantage to (instrumental) resource exploitation by turning nature preservation into a peculiar, and largely indefensible, special case of resource exploitation and consumption. Maintaining the distinction between intrinsic and instrumental value, in contrast, allows us to set certain things aside and exempt them from use. Because the instrumentalist approach to valuing natural objects is the primary approach in economics, the valuable contribution that can be made by an intrinsic value approach has been neglected. In the nineteenth century, nevertheless, newly formed national parks were valued in two ways: in terms of their use as sources of raw materials and in terms of their intrinsic value as aesthetically and scientifically interesting collections of natural objects. Proponents of these parks argued that these geographical areas, though instrumentally worthless (Yellowstone was said to have no minerals worth extracting and to be unsuited for agriculture), were priceless (off the monetary scale) aesthetically and scientifically. In these arguments, intrinsic and instrumental values were juxtaposed against each other as competing perspectives, both of which could to some degree be expressed in economic terms. Interestingly, however, arguments in terms of intrinsic value were always estimated in extreme terms: the natural objects were declared to be priceless, or off the economic scale, too valuable for any reasonable price to be set upon them. What we have here is an attempt to produce the trumping effect of the rights for nature arguments of the twentieth century, without corresponding theoretical problems. All that is being said is that these areas are valuable in a noninstrumental way such that they should be removed from the market system, specifically, instrumental resource exploitation. The judgmental process is identical to the one that excludes the ornamental knife from use. In effect, the assignment of aesthetic and scientific intrinsic value to these natural areas is an attempt to give them the status of economic externalities, providing a useful tie with standard economic valuational theory. The public incurs the social costs involved in not exploiting these natural areas in much the same way that it incurs the social costs involved in exploiting other parts of nature, for example, covering the cost of pollution not included directly in the cost of the manufacturing of specific products.[40]

In part, to be sure, the high economic valuations (that they are priceless) are sometimes produced, as a matter of strategy (when the valuations are inflated), to override any possible instrumental argument. Nevertheless, they generally reflect people's basic evaluational intuitions, and can usually be justified (provided they are in accordance with social standards and ideals) on the grounds that their economic value (from an aesthetic/scientific intrinsic value perspective), though high, is indeterminate and speculative, given that the routine economic transactions that determine worth in the market system do not occur (or at least seldom occur) with regard to natural objects.[41] Recent sales of houses in a particular area, for example, largely determine the value of houses in future sales. Because natural areas (and species) are not bought and sold like houses on an everyday basis, their value cannot be determined in this way. To come close to this kind of valuing, we have to ask ordinary people to engage in contrary-to-fact thought experiments in which they imagine what dollar values might be assigned if natural areas and species were bought and sold like houses and if there actually were large numbers of humans who traded in them on a regular (and quotable) basis. This problem is analogous to the problem of determining the value of paintings and other art objects, most of which are also not bought and sold in the marketplace (having been removed from the market system by being placed in publicly supported museums). The value of the Mona Lisa, though very high (essentially priceless), is indeterminate and speculative because there is nothing comparable to housing sales to bracket the price range, which is another way of saying that the Mona Lisa, and similar paintings, are external to the market system that determines the value of less highly valued paintings and reproductions. Note that this valuing is not mystical or confusing. It clearly reflects our desire as individuals, as a society, and as a historically evolved culture to value some things noninstrumentally and to set them aside and protect them from exploitation. It is justified by the fact that the valuational methods of the market system are designed to provide values for things that are in the market and subject to market forces, not for things that have been removed from the market and are, therefore, external to it.

Anthropocentrism revisited

I HAVE USED the term *weak anthropocentrism*, rather than simply *anthropocentrism*, in the title of this paper to help call attention to the fact that not all anthropocentric valuing is instrumental. Without the addition of the word *weak*, no doubt many nonanthropocentrists would probably conclude that the title contained a typographical error or was a contradiction in terms: "instrumental intrinsic value." While I do not think that labels are important, it is useful to call the view I represent weak anthropocentrism at least until it becomes generally recognized that anthropocentrism does not imply instrumentalism. I do not think that it is possible for humans to avoid being anthropocentric, given that whatever we humans value will always be from a human (or anthropocentric) point of view. Even when we try to imagine what it might be like to have the point of view of (or be) a bat, a tree, or a mountain, in my view, we are still looking at the world anthropocentrically, the way a human imagines that a nonhuman might look at the world.

This kind of anthropocentrism, as I noted above, is built into Callicott's anthropogenic position that "the *source* of all value is human consciousness" and Rolston's aesthetic position that "the experience of beauty is something that humans bring into the world."[42] Although Rolston tries to deemphasize such human values, covering beauty almost as an afterthought after elaborately developing his nonanthropocentric value theory, these values do play a major role in his environmental ethics. For example, as Rolston acknowledges at the end of his chapter on the objective intrinsic value of organisms, the good of human (anthropocentric) aesthetic appreciation can easily override the good of its own of a wildflower:

> The goods preserved by the human destruction of plants must outweigh the goods of the organisms destroyed; thus, to be justified in picking flowers for a bouquet one would have to judge correctly that the aesthetic appreciation of the bouquet outweighed the goods of the flowers destroyed. One might pluck flowers for a bouquet but refuse to uproot the whole plant, or pick common flowers (daisies) and refuse to pick rare ones (trailing arbutus) or those that reproduce slowly (wild orchids).[43]

In this example, the deciding factors are purely anthropocentric: the human is the judge and the issue is the amount of aesthetic pleasure the human believes he or she will receive by destroying the nonanthropocentric good of the plant.

Rolston's nonanthropocentrism is also infected with anthropocentrism in an even more fundamental sense, for he argues, against the biocentrism of Taylor, that humans are superior to the rest of nature and deserve special consideration, a *strong* anthropocentric claim.[44] The practical effect of this position is an anthropocentric point of view in which humans receive special or superior attention because of their special or superior status. Rolston's arguments that nonhuman organisms have nonanthropocentric goods of their own have no *practical* impact on this anthropocentrism, which retains its traditionally anthropocentric character, as a kind of stewardship, rather than as a form of dominion.[45] (It is probably because humans are superior to wildflowers that picking them for the aesthetic enjoyment of humans can be justified.)

The term *nonanthropocentric intrinsic value* is really more problematic than the term *anthropocentric intrinsic value*, for the former comes very close to being redundant. The word *intrinsic*, which here is supposed to mean "for its own sake," clearly distinguishes the value of the valuable thing from the value of the valuer, indicating that the value is tied to the sake of the thing said to be intrinsically valued, not to the sake of the valuer. At best, the word *nonanthropocentric*, which is supposed to refer to value that is not human centered and dependent on human judgment, merely asserts that the value of the thing valued does not derive its value from the value of a human valuer, a point already made more generally (with regard to any valuer) by the term *intrinsic*. If we come to accept, as I have argued above, that most values are independent of human judgment, and that when we do value, we value necessarily from a human perspective, but not necessarily in terms of human instrumental interests, then the term *intrinsic value* (referring both to intrinsic value conferred through human judgment and intrinsic value defined as the goods of their own of living organisms), will make the terms *nonanthropocentric* and *anthropocentric* unnecessary.

NOTES

1. Aldo Leopold, "The Land Ethic," in *A Sand County Almanac, and Sketches Here and There* (New York: Oxford University Press, 1949), 209–10.

2. There are two references to rights of nature in Aldo Leopold's "The Land Ethic," *Sand County Almanac*, 204, 211. Environmentalists were also influenced by Christopher Stone, *Should Trees Have Standing? Toward Legal Rights for Natural Objects* (Los Altos: William Kaufman, 1972), which, though not focused on moral rights for nature, left open the possibility. Because animal liberation and environmental ethics were ambiguously interrelated in the early seventies, calls for rights for domestic animals were also frequently considered to be calls for rights for nature. Rights for animals

were being championed at that time by Peter Singer, *Animal Liberation: A New Ethic for Our Treatment of Animals* (New York: New York Review/ Random House, 1975). Rights talk was so common among environmentalists by the mid-seventies that John Passmore considered it to be the basic preservationist position. See John Passmore, *Man's Responsibility for Nature: Ecological Problems and Western Traditions* (London: Duckworth, 1974), 115–17. See also Norman Myers, *The Sinking Ark* (Oxford: Pergamon Press, 1979), 46; David Ehrenfeld, *The Arrogance of Humanism* (New York: Oxford University Press, 1978), 207–09; Paul and Anne Ehrlich, *Extinction* (New York: Random House, 1981), 48. For detailed discussion of people holding this view at the movement level, see Roderick Frazier Nash, *The Rights of Nature: A History of Environmental Ethics* (Madison: University of Wisconsin Press, 1989).

3. Discussion of rights for nature was a basic theme of many issues of *Environmental Ethics* in the first five years.

4. For example, according to Samuel Alexander, beauty, whether artistic or natural, is contemplated for its own sake, that is, is regarded as being intrinsically valuable. Samuel Alexander, *Beauty and Other Forms of Value* (London: Macmillan and Co., 1933), 13–14. The idea that art objects are intrinsically valuable is so well established that it was seldom overtly expressed. It appears primarily in analogies extending intrinsic value from art to nature and in contrasts of intrinsic value with instrumental and utilitarian value. See, for example, Thomas Cole, "Essay on American Scenery," in John W. McCoubrey, ed., *American Art: 1700-1960* (Englewood Cliff, N.J.: Prentice-Hall, 1965), 99–100. According to Emerson, "art aims at beauty as an end," that is, as an intrinsic rather than an instrumental value, and beauty, both in nature and in art, "gives us delight *in and for themselves.*" See Ralph Waldo Emerson, "Thoughts on Art," in McCoubrey, *American Art*, 76, and "Nature," in *The Writings of Ralph Waldo Emerson*, ed. Brooks Atkinson (New York: Modern Library, 1940), 9. At the beginning of this century, G. E. Moore argued that things that are beautiful in art and nature are good for their own sakes and "the proper appreciation of a beautiful object is a good thing in itself" and he compared landscape paintings with natural landscapes. G. E. Moore, *Principia Ethica* (Cambridge: At the University Press, 1903), 188–89. In accordance with this tradition, Aldo Leopold draws similar comparisons in "The Conservation Esthetic," in *Sand County Almanac*, 168, and "Means and Ends in Wild Life Management," *Environmental Ethics* 12 (1990): 330.

5. Although this definition is implicit in virtually all the writings of the deep ecologists, it has been explicitly stated by J. Baird Callicott (who is not a deep ecologist): "An anthropocentric value theory (or axiology), by common consensus, confers intrinsic value on human beings and regards all other things, including other forms of life, as being only instrumentally valuable, i.e., valuable only to the extent that they are means or instruments which may serve human beings. A non-anthropocentric value theory (or axiology), on the other hand, would confer intrinsic value on some non-human beings." J. Baird Callicott, "Non-Anthropocentric Value Theory and Environmental Ethics," *American Philosophical Quarterly* 21 (1984): 299.

6. See Henryk Skolimowski, "The Dogma of Anti-Anthropocentrism and Ecophilosophy," *Environmental Ethics* 6 (1984): 283–88.

7. See, for example, Monroe Beardsley, "Intrinsic Value," in Monroe Beardsley, *The Aesthetic Point of View: Selected Essays*, ed. Michael J. Wren and Donald M. Callen (Ithaca and London: Cornell University Press, 1982). After applauding Dewey's attack on intrinsic value and adding arguments of his own, Beardsley admits in his concluding remarks that he himself finds Dewey's attack confusing: "I am always frustrated in reading Dewey, trying to separate the enormously good points from the confusing ones. Much of Dewey's famous attack on intrinsic value is really concerned with something else, namely ends-in-themselves (as opposed to ends-in-view). What he really exposes over and over again is the danger of fixing goals without reasonable regard to their means and consequences, and he is convinced that the belief in intrinsic value fosters this fixation, with its attendant train of ills: fanaticism, utopianism, opportunism, and the rest. Of course, it does not logically follow that if there are intrinsically valuable things, then there are necessarily ends-in-themselves" (63). Other frustrations and confusions are not difficult to find. For example, in speaking of the consummatory phase of experience (happily, consummation, not consumption), Dewey warns against "the joys of egotistic success" replacing "the fulfillment of an experience for its own sake." He then claims to solve this problem by expanding the definition of "instrumental" (his quotation marks) to cover such matters, in passing, scolding "persons who draw back at the mention of 'instrumental' in connection with art" (that is, who hold that aesthetic experiences involve intrinsic value, not just expanded instrumental value). He fails, however, fully to explain how the quotation marks take care of the problem of intrinsic value. John Dewey, *Art as Experience* (New York: G. P. Putnam and Son, 1934), 139. Amazingly, Beardsley discusses Dewey's attack on intrinsic value without ever bothering to cite any book or essay by Dewey on the subject, leaving uninformed and skeptical readers like myself no opportunity to judge for themselves whether or not the attack was truly successful. The difficulties involved in finding and unscrambling Dewey's attack suggest that it was more an ideological pronouncement than a real argument.

8. See A. J. Ayer, "Critique of Ethics and Theology," in *Language, Truth and Logic* (New York: Dover Publications, 1950), 102–20.

10. Eugene C. Hargrove, "The Role of Rules in Ethical Decision Making," *Inquiry* 28 (1985): 3–42.

11. Ibid., 26.

12. Ibid., 27.

13. See, for example, Paul W. Taylor, "Are Humans Superior to Animals and Plants?" *Environmental Ethics* 6 (1984): 151, n. 5.

14. Hargrove, "Role of Rules," 10–11. I go on to argue that the treatment of nonconstitutive rules in a constitutive manner in chess and ethics limits a player's or moral agent's perception (in an Aristotlian sense), causing the player to miss winning opportunities and the moral agent to overlook important aspects of and solutions to moral problems (18–23).

15. As John Passmore once noted, in criticism of environmental ethics theory in general, "It is a considerable presumption, indeed, that our descendants

will continue to admire wildernesses aesthetically, just as it is a considerable assumption that they will continue to enjoy solitude. . . . We ought . . . to preserve wildernesses because they may turn out to be useful and because they may afford recreational pleasures, scientific opportunities and aesthetic delight, to our successors. The first of these considerations . . . is a powerful one, the others less powerful in that they rest upon the presumption that our descendants will still delight in what now delights only some of us and did not delight our predecessors." Passmore, *Man's Responsibility*, 109–10.

16. Paul W. Taylor, *Respect for Nature: A Theory of Environmental Ethics* (Princeton: Princeton University Press, 1986); Holmes Rolston, III, *Environmental Ethics: Duties to and Values in Nature* (Philadelphia: Temple University Press, 1988).

17. The issue is apparently what Callicott calls "moral monism." See J. Baird Callicott, "The Case against Moral Pluralism," *Environmental Ethics* 12 (1990): 99–124. Callicott classifies Christopher Stone and myself, among others, as moral pluralists and cites himself and Holmes Rolston, III as moral monists.

18. William Frankena lists utility values, extrinsic values, contributory values, and perhaps inherent values, and other formulations are possible. William Frankena, *Ethics*, 2d ed. (Englewood Cliffs, N.J.: Prentice-Hall, 1973).

19. Paul W. Taylor, "Are Humans Superior to Animals and Plants?" 150–51.

20. Ibid., 151–52.

21. See Eugene C. Hargrove, *Foundations of Environmental Ethics*, (Englewood Cliffs, N.J.: Prentice-Hall, 1989), 124–29; 165–66; 208–09.

22. See Bryan G. Norton, "Environmental Ethics and Weak Anthropocentrism," *Environmental Ethics* 6 (1984): 133–38. Although my remarks here are in the spirit of Norton's paper, we differ concerning the meaning of weak anthropocentrism (personal communication).

23. I develop this point at great length in my book, *Foundations of Environmental Ethics*.

24. Holmes Rolston, III, *Environmental Ethics* (Philadelphia: Temple University Press, 1988), 97–104, 169, 186–87, 232–45.

25. Ibid., 169.

26. Ibid., 218.

27. Ibid., 187. Apparently the claim is that anything that is instrumentally valuable to a holder of intrinsic value, an organism, in some way shares in the intrinsic value of that organism.

28. Ibid., 233–45.

29. J. Baird Callicott, *In Defense of the Land Ethic* (Albany, N.Y.: SUNY Press, 1989), 173.

30. I use the word *uncritical* to describe the claim that humans have intrinsic value because Callicott introduces the claim on the basis of the following sentence: "The intrinsic value of oneself has *for some reason* been taken for granted. . . ." Ibid., 172 (emphasis added).

31. I am aware that deep ecologists try to overcome this problem by taking a Spinozistic approach. See George Sessions, "Spinoza and Jeffers on

Man in Nature," *Inquiry* 20 (1977): 481-528. I do not, however, see any evidence that Callicott is a Spinozan.

32. Ibid., 147.

33. Ibid., 133–34.

34. See Hargrove, "The Role of Rules," 21 and 22, 40.

35. See, for example, Anthony Weston, "Beyond Intrinsic Value: Pragmatism in Environmental Ethics," *Environmental Ethics* 7 (1985): 321-339, who relies on Beardsley, "Intrinsic Value," cited above, and Bryan G. Norton, "Environmental Ethics and Weak Anthropocentrism," *Environmental Ethics* 6 (1984): 131–148. Norton himself (personal communication) backed into pragmatism as a result of the ontological complications of nonanthropocentric intrinsic value, but did not fully embrace pragmatism until he wrote "Conservation and Preservation: A Conceptual Rehabilitation," *Environmental Ethics* 8 (1986): 195–220. There are, of course, also anti-environmental pragmatic instrumentalists. Passmore, *Man's Responsibility for Nature,* for example, argues that nature preservation can only be supported instrumentally (101–26), claiming eventually that "whatever exists in nature is of some use to [humans]." He notes that though this claim is not "an empirical hypothesis, for there is no way to falsify it," it is, nevertheless, a useful "guiding principle," which should not be "set aside" (180).

36. Norton, "Environmental Ethics and Weak Anthropocentrism."

37. See Nash, *Rights of Nature.* See also F. Fraser Darling, "Man's Responsibility for the Environment," in F. J. Ebling, ed., *Biology and Ethics,* Symposia of the Institute of Biology, no. 18 (London, 1969), 119.

38. For a more detailed discussion, see Hargrove, *Foundations of Environmental Ethics,* 124–29.

39. Rolston, *Environmental Ethics,* 16–17, 25–26.

40. For a longer discussion of this matter, see Hargrove, *Foundations of Environmental Ethics,* 210–14.

41. While there are undoubtedly millions of people who have sold a piece of land with trees on it, there are probably only a few hundred people who have bought or sold a forest. Although forests are apparently bought and sold all the time, I have never had regular contact at any time in my life with anyone who has either bought or sold a forest. I would guess that the number of people who have consciously bought or sold a species is less than one hundred, if that many.

42. Callicott, *In Defense of the Land Ethic,* 133; Rolston, *Environmental Ethics,* 233.

43. Rolston, *Environmental Ethics,* 120.

44. Ibid., 62–78.

45. Anthropocentrism of this kind has even appeared in the more recent writings of Callicott, who also gives special attention to humans over domestic animals and to domestic animals over wild ones. See Callicott, "Animal Liberation and Environmental Ethics: Back Together Again," in *In Defense of the Land Ethic.*

*
**

The Future of Ecology

Michael E. Zimmerman

ADDRESSING THE "FUTURE" of ecology is a risky business. The risks involved here are not merely a matter of making mistakes in predicting the future of a complex science and of a social movement whose goal is to convince people to act in ways consistent with an ecologically-informed understanding of the ecosphere. In addition to the complexities of forecasting, we are faced with a more subtle problem when speaking of the future of ecology. This problem concerns the fact that in speaking of a future which will be improved by dramatic human action as informed by scientific findings, ecological activists adopt central features of the progressive notion of the future which is arguably responsible for our present ecological crisis. In this essay, I will examine the extent to which *radical ecologists* unwittingly reproduce the modern impulse toward total control, an impulse of which they are usually quite critical.

The progressive view of the future has not always characterized humankind's understanding of its authentic possibilities. For example, traditional peoples regarded the future as a variation on an ancient cycle of events, a cycle shaped by local and climatic conditions. The future was not viewed as an unpredictable and radical departure from the tribe's customary practices, but instead as a dimension of the self-enclosed circle of mythical time. The myths of such traditional peoples are conservative. Their myths emphasize conformity to sacred law and sacred time. Such conformity often proved consistent with the long term well-being of the habitat of such peoples. Taboos against hunting certain animals during particular seasons, for instance, may have (more than accidentally) coincided with reproductive habits of the deer population.

In the Western world, Jews and Christians broke out of the cyclical conception of time. They portrayed history as involving a series of unique events, directed by Divine providence, and moving toward a glorious and peaceful new era. In this linear view of history, the future was depicted as the time of radical change that would transform a world currently laden with suffering and sin. Over the centuries, Europeans gradually began conceiving of the future in secular terms. During the Renaissance, for instance, people began suggesting that Divine providence was not the only active ingredient in shaping the future; humankind might also play an important role in this matter. Enlightenment thinkers completed the secularization process and ushered in the period I will call "modernity," the optimistic age in which humankind seeks to gain total control of its destiny by applying scientific and technological skills to mastering the conditions of life. For modernity, according to its progressive ideologues, Marxist and liberal alike, the future involves constant improvement in the human estate, improvement owed to increasing gains in humankind's capacity to control nature. Faith in this secular ideal of progress became very strong in the late nineteenth century.

Events of the twentieth century undermined much of such faith in the linear progress promised by modernity. The Somme, Auschwitz, and Hiroshima; Bhopal, Chernobyl, and the burning oil fields in Kuwait—recalling these familiar names is enough to cast a pall on any naive optimism about the inevitable improvement of human life. In the past twenty years the general public has become aware of the possibility that global catastrophe might result not only from the use of atomic weapons invented by industrial technology, but also from ecological decay brought about by the rampant growth of that very technology. The audacious attempt to "dominate" nature may backfire. Many critics argue that the meliorist concept of a progressive future, when coupled with an arrogant anthropocentrism, inevitably leads to exploitative treatment of the ecosphere, treatment that might prove suicidal for our species.

In the face of problems such as global warming, acid rain, desertification, loss of tropic forests, and thinning of the ozone layer, important debates are now taking place concerning how to cope with such planetary ecological dysfunctions. In the 1990s ecological issues will play a central role in deliberations on economic, political, religious, and personal affairs. Predictions about how all these debates and discussions will turn out is what we normally mean by "the future of ecology." But what kind of ecological thinking will predominate in such deliberations? Will we simply attempt to reform current practices without calling into question the anthropocentrism which seems to play an important role in generating ecological problems, or will we begin to question more deeply the philosophical roots of anthropocentrism? Can we retain what is valid and important from the "progressive" view of history, without simultaneously justifying the ethnocentric and anthropocentric attitudes arguably responsible for our twentieth-century social and ecological crises? Can we develop an interpretation of human history which retains what is valid from our cultural tradition concerning humanity's place on Earth while replacing what is invalid with new philosophies and values? Although such questioning has begun to take place, it would be naive to expect that a "deeper" approach to ecological concerns will be entirely free of the problematic attitudes which originally gave rise to ecological problems.

Those of us who are concerned about ecological problems are very much a part of modern civilization. Hence, we very often take an activist stance toward the "future" of ecology. We want the future to be better than it is today. Some of us even have utopian visions of a far smaller population of humanity dwelling peacefully on an Earth covered with forests and grasslands, all of which are permeated by many diverse forms of life. Even ecological activists with a less sanguine view of the future adopt the attitude central to modern political ideologies: that humanity can take the future into its own hands and shape it for the better in accordance with our growing scientific knowledge. Rather than allowing the allegedly suicidal plunder of the planet to continue unabated, ecological activists want to intervene in history in order to produce a future conducive to the well-being of all life on Earth. They call for a vastly reduced human population, much more free land for wildlife, a socially richer but materially simpler lifestyle for people in wealthy countries, a better material standard for people in currently poor countries, cleaner and more efficient production processes, political units centered in bioregions, and a shift from anthropocentrism toward an attitude of respect for the whole ecosphere on which all life depends.

There is much in this vision of the future with which I can agree. The question is, however, whether such a future—utopian by some standards—can ever be achieved by an activism which at times seems to share the goal-directed, future-oriented attitudes which brought us the ecological crisis in the first place. Does ecological activism, quite unintentionally, risk reproducing some of the same problematic social, political, and even ecological consequences which we associate with other modern activist movements which have attempted to control the future? Do all attempts to shape the future on a global basis inevitably cause greater problems than the ones being addressed in first place?

Debates about these questions are charged, even within the circle of radical ecologists. Consider, for example, an exchange that occurred in August, 1989 at "The Wilderness Condition" conference, organized by Max Oelschlaeger in Estes Park in the Colorado Rockies. At the conference, a leading deep ecologist, George Sessions, presented a paper called "Ecocentrism, Wilderness, and Global Ecosystem Protection." Sessions argued that non-stop industrial "development" and overpopulation in Third World countries was pushing wild nature to the point of extinction. At current logging rates, for example, most of the vast Amazonian rainforest will disappear in only thirty years. To deal with such vast threats not only to wild nature but to the ecosystem as a whole, Sessions called for the development of new political structures and bureaucratic organizations. Sessions remarks:

> Increasingly, our environmental problems are being recognized as global in scope and, as such, require effective international cooperation. The United Nations also needs to reorganize its population control agencies and environmental protection programs to reflect a unified Ecosystem Protection approach. Ecosystem and environmental protection must be given a very high priority on its agenda.[1]

In answering a question about his paper, Sessions—reflecting the urgency which many people now feel in the face of such widespread destruction of wild nature—emphasized his point that time is running so short that governments must intervene to prevent the final destruction of remaining wild places. At this point, Gary Snyder objected. Snyder is well-known for stressing the importance of bioregionalism, for promoting the well-being of Turtle Island (North America), and for his attempts to practice his deep ecological attitudes by living simply in land high in northern California's Sierra Nevada mountain range. Although a friend of Sessions and appreciative of his work in deep ecology, Snyder expressed dismay that George Sessions was appealing to big government to intervene on a planetary scale to deal with ecological problems. From Snyder's viewpoint, big government was itself so complicit in the current ecological crisis that calling on such government for help was akin to inviting the fox to guard the chicken coop. In view of the at best spotty accomplishments of the U.S. Environmental Protection Agency's efforts to curb industrial pollution, and in view of how the U.S. Forest Service finances the clear-cutting of national forests at well below market value, we may well understand the reasons for Snyder's skepticism about Sessions' proposal for international governmental intervention on behalf of wilderness. In reply to Sessions' query about the alternative, Snyder insisted that there is no real alternative to individuals acting at the grass roots level to effect changes in their own immediate locales. Big government won't be able to save us, for big government is part of the problem.

Because George Sessions is a deep ecologist, he is well-informed about the advantages of bioregionalism and the problems of big, centralized government. He has called for new forms of political association which move away from the centralization and hierarchy so characteristic of the industrialized nation states. Nevertheless, he is plainly skeptical about whether a sufficient number of movements can arise at the grass roots level and whether they will rise swiftly enough around the world to make the large scale changes necessary to save wilderness from total elimination and to save the ecosphere from potentially irreversible harm.

The debate between Sessions and Snyder has been reproduced in a somewhat different form in an exchange between

Wendell Berry and Alan Atkisson. Berry, the noted Kentucky farmer, poet, and author of books such as *The Unsettling of America* and *The Gift of Good Land*, has criticized the possibility of acting on a planetary scale to save the ecosphere.[2] In a 1989 commencement address, Berry remarked that everything important involves relations with what is local and concrete. Successful relationships presuppose personal, subtle knowledge of those people and those things with whom we are in relationship. When people resort to abstractions to comprehend their situation, authentic relationships can't be sustained.[3] One does not converse with people on a personal level—at the level of genuine relationship—in terms of the abstract language of social causes, such as the civil rights movement or the women's movement, though such rhetoric initially may have been helpful for personal relations. Hence, when representatives of the ecological movement begin to use the term "planetary," Berry shudders, for this indicates that even ecologists have forgotten the importance of the concrete and have been seduced by the sirens of the abstract. Acknowledging that concern for the "planetary" expresses the fact that no single place can be fully healthy until all others are, Berry nevertheless claims that it is "preposterous" for anyone to think that he or she can "heal the planet." From Berry's viewpoint, there are in fact no planetary problems, only local ones; hence, there are no "large scale solutions" for there are no "large scale problems."

Our problem is that we are living wrongly, personally and communally. Thus, we must devise new economic methods for sustaining our lives, methods which emphasize diagnosis and solution of local problems. Supposedly, just as the civil rights movement has not given us better communities, so too the environmental movement has not given us a better relationship to nature. Berry asserts that the question "is not how to care for the planet but how to care for each of the planet's millions of human and natural neighborhoods, each of its millions of small pieces and parcels of land, each one of which is in some precious way different from all others."[4]

Only love, which is never abstract, can establish the relationship with a particular place needed to heal that place. Such love is necessarily absent in large scale organizations and bureaucracies. We allow them to continue to wreak widespread ecological destruction precisely because we are not really willing to live the simpler, poorer lives that would be required if we stopped our parasitical treatment of nature and poor people. The horrifying ecological "accidents" of the industrial age, such as Chernobyl and Bhopal, were direct consequences of our attempts to act beyond appropriate limits. The dreadful consequences of such hubristic acts are "revenges of Nature." Only by humbly accepting our limited role in a particular place on Earth can we learn to live in accordance with the limits imposed by Nature.

Alan Atkisson, executive editor of the notable journal, *In Context*, has offered a sympathetic and insightful critique of Berry's position. While sharing Berry's conviction that local action is vital, and while acknowledging the fact that we don't have the competence to "manage planets," Atkisson nevertheless argues that we do in fact have "planetary problems."[5] Only by addressing such problems appropriately can we eventually arrive at the sort of community for which Wendell Berry longs. It is true that all our ecological problems can be traced back to the action of specific individuals.

> But understanding that simple fact—crucial to empowering us to act—does not erase the reality of larger structures, both natural and human-created. For the effects of these problems are global. Many of the systems in which these individuals and households are embedded are global. Some of the individuals making those decisions and demands are controlling organizations whose reach, indeed whose very identity, is often global. To ignore this is to misrepresent the reality—a *global* reality, that is, admittedly hugely complex and beyond the ability of any one mind to comprehend fully. But that does not mean we should abandon the attempt.[6]

Moreover, Atkisson maintains that the language of civil rights and women's rights has in fact been beneficial to communal and social relationships, just as the Endangered Species Act has proved helpful to protecting rare flora and fauna. That the success of these "abstract" ideas has not been greater can be attributed to the enormous inertia of those social structures responsible for problems in the areas of race, gender, and environment. The idea of "loving the planet" retains its invigorating power. Indeed, Atkisson tells us, for the city dweller who lacks the caring proximity to land enjoyed by Berry, loving the planet (and thus "thinking globally") may be a crucial motivation for learning to "act locally" in an ecologically responsible manner. Unlike Berry, Atkisson understands that as the Earth becomes girdled by electronic media, people from all parts of the globe can be in instantaneous contact—and this can affect events on a global basis. On our shrinking planet, we *must* learn to think globally if we are to act effectively locally.

In a reply to Atkisson, Berry reiterates his belief that it is simply impossible to "think globally."[7] Instead, one can only think and act locally. A good action has to be "scaled and designed so that it fits harmoniously into the natural conditions." Denying that he finds any inspiration in photographs of the Earth taken from satellites, Berry claims that those images remind him of a Christmas ornament. Only his own neighborhood—"large, mysterious, painful, joyful, and lovely"—can inspire him to act appropriately. For Atkisson, as for many others working with ecological problems on the local, national, and global scales, the same images of the planet inspire not an attitude of abstraction and indifference, but rather a profound sense of appreciation for the beauty and complexity of planet Earth.[8] Out of such appreciation, he believes, there can emerge a love that is both oceanic and tangible, global and local—a love that can inspire appropriate action at various levels.

The debates between Sessions and Snyder, and Atkisson and Berry are in some respects reminiscent of the currently heated debate between modernists and those calling themselves postmodernists. (Of course, both Sessions and Atkisson are largely critical of modernity, for reasons in many ways similar to those advanced by postmodern thinkers. Nevertheless, they continue to recognize that attempts to conceptualize our problems on a large scale, even on a planetary scale, are important, though they also caution that the ecosphere is far too complex ever to be fully understood.) Optimistic Enlightenment thinkers believed that abstract and universal concepts would make it possible both to gain control over nature and to pacify an unruly humankind in ways that would vastly improve the human estate. Liberal capitalism and especially Marxism developed totalizing narratives of human emancipation, narratives which depicted as reactionary or useless those forms of social life which differed from the rise of universal humanity and thus impeded its progressive unfolding. Capitalists and Marxists alike developed the view that the *telos* of world history was leading to the realization on a planetary basis of their "universal" conceptions of humanity. Armed with almost magical technological skills, either the proletariat or the bourgeoisie would usher in a glorious new age in which "man" would finally become master of his own fate.

Postmodern thinkers, such as Lyotard, Derrida, Foucault, and Rorty display "incredulity" toward such totalizing narratives. While promising emancipation, these narratives have more often brought complex new forms of oppression and the untold horrors of modern war. Marxism led to the Gulag, while liberal capitalism led to colonialism and imperialism. In their titanic Cold War struggle, state communism and monopoly capitalism turned Third World countries into pawns, thereby helping to destroy their people and lands. Under the banner of "universal human freedom," powerful First World countries march across the planet, compelling Third World peoples to abandon traditional ways of life and to sacrifice their lands in order to meet the insatiable demands of the global market.

In view of all this, we may well understand the skepticism voiced by Gary Snyder and Wendell Berry regarding the wisdom either of forming supranational governmental agencies to administer land use on a planetary basis, or of attempting to "think globally" in order to save the planet. Postmodern critics share such skepticism. They might argue that deep ecology itself, despite its critique of the ecological destructiveness of modernity, has internalized its totalizing impulse. While postmodernists and deep ecologists alike are critical of anthropocentric humanism, postmodernists might wonder whether deep ecologists tend to elevate biocentrism into places vacated by humanism. While earlier in this century, draconian political measures were taken in order to further the lofty ideals of "humanism," at the start of the next century might equally repressive political measures be taken in order to further the lofty ideals of "ecocentrism"? In other words, does deep ecology, perhaps unwittingly, foster a new form of totalitarianism, a kind of ecofascism?

The postmodern ecological thinker, Jim Cheney, has argued that deep ecologists do in fact offer a totalizing vision of the world, a metaphysical vision inspired by their sense of isolation and alienation.[9] Like the ancient Stoics, deep ecologists preach the gospel of conformity to cosmic law as embodied in the workings of the ecosphere. Just as the Stoics were motivated to seek a cosmic home because of the increasing disorder and fragmentation of the Roman Empire, so too deep ecologists look for a higher cosmic order with which to identify because of the collapse of the anthropocentrism of modernity. For deep ecologists the achievement of cosmic unity involves the expansion of the self's capacity for identification. The wider the field of that with which one can identify, the more one moves toward self-realization, understood as the experiential and metaphysical actualization of Atman or Self that permeates the whole cosmos. Leaving behind the confines of the paranoid ego-subject, the deep ecologist expands his "self" to the level of the cosmic "Self," Atman. Wider identification leads us to embrace what we formerly regarded as other—as something to be dominated, for example. Hence, the Australian deep ecologist John Seed claims that when a person can intuit that "I am the rainforest," he or she will act spontaneously to preserve it, just as he or she would act spontaneously to take care of part of himself or herself that was threatened.

While Cheney largely agrees with deep ecology's critique of anthropocentrism and with the premise that nature is intrinsically valuable, he suspects that deep ecologists have failed to discern the totalizing—and thus the modernist—implications of their views about Self-realization. The Stoics of old were faced with the disappearance of the unifying force of the Empire. Cheney remarks that

What has disappeared for *us*, creating the vacuum in which Ecosophy S has appeared, is the self-congratulatory security of *modernism* and (most important for deep ecologists) its declared anthropocentricity. [Deep ecology] . . . functions in a manner analogous to Stoicism, but within a different context, that of the demise of modernism, its shattering into a world of difference, the postmodern world. It expresses a yearning for embedment coupled with a refusal to forgo the ultimate hegemony so characteristic of modernism. This is the *subtext* underlying the concepts of Self-realization, identification, and ecological consciousness. Subtextually, the central operative idea at work in these concepts is the idea of *containment*, containment of the other, of difference, rather than the genuine *recognition* of the other, genuine acknowledgment and embracing of the other.[10]

For Cheney, the Self-realization espoused by deep ecology is bound up with a masculinist sense of self unable to tolerate difference. Despite the rhetoric of "letting things be," then, deep ecologists yearn for a lost primal unity and seek to regain it by incorporating everything within themselves. In their talk about Self-realization through wider identification, deep ecologists are in fact simply calling on us to expand the alienated masculinist ego to the cosmic level. By way of contrast to this project, Cheney recommends that radical environmentalists take the risk of opening themselves to otherness, instead of trying to contain it within themselves. Such an opening would involve the possibility of developing a different sense of self, not one which gained security by swallowing or incorporating everything, but one which experienced the joy and sorrow of concrete relationships with people and things that are different, irreducible to one's own customary self-understanding.

To some extent this is the path recommended by Snyder and Berry, though Cheney might discern in their writings subtexts of modernity equally problematic as those he purports to have revealed in the writings of deep ecology. Cheney's criticism of deep ecology is important for reasons analogous to why the skepticism of Snyder and Berry regarding "planetary ecological bureaucracies and "global thinking" is important. We must be wary of grandiose schemes purporting to save us, for they may well end up forcing everyone and everything to conform to a one-dimensional and totalitarian conceptual, political, or economic grid. We must learn to pay attention to our own backyard, to question our received understanding of "self" and "world," to respect and to celebrate difference, to develop relationships with precisely that which most challenges our identity—relationships which let the other be what it is, without my "identifying" with it in such a way that its otherness and its difference somehow get eclipsed in the process.

Nevertheless, I also share with George Sessions the conviction that current conditions do require that we begin to organize ourselves at a planetary level, because our species is already acting in ways that have planetary consequences. Moreover, I share with Alan Atkisson the experience of wonder, awe, and appreciation brought on by those photographs of Earth from outer space, just as I experience awe in the face of beautiful sunsets over the Mississippi River or when watching a blue jay hop purposively around in my back yard. I also share with deep ecologists the sense that wider identification with things may be an important step toward overcoming the subject-object dualism and the anthropocentrism which are such important sources for today's ecological crisis.

Cheney presupposes that the sense of "Self" in Self-realization is simply an expanded version of the masculinist sense of self—that which fears otherness and seeks to eliminate it. In fact, as I have argued elsewhere,[11] greater identification may also be understood as breaking apart the centered ego-subject and leading to deeper relationships at the local level as well as at larger levels. Looking with suspicion, as Cheney does, on the deep ecologists' yearning for a sense of unity with the cosmos is reminiscent of the suspicion Freud expressed regarding a friend's report of the "oceanic feeling" of oneness with the universe. Freud never had such an experience, so he tended to interpret it in terms he could understand: as the yearning of the ego to regress to the blissful infantile state of total absorption in its mother. Cheney's hostility toward the yearning for unity leads him to emphasize the privileged status of difference. But by creating such an opposition, Cheney himself (perhaps unintentionally) reproduces a version of the dualism that is so central to modernity, the binary dualism which elevates one pole to a position of superiority at the expense of the other.

In speaking of wider identification, deep ecologists are in fact continuing in somewhat different guise an essential ingredient of Enlightenment modernity's project of emancipation. Deep ecologists want to free nature from needless oppression, just as abolitionists wanted to free the slaves. Heirs of the

Enlightenment sought to overcome intolerance, bred by fear and hostility toward difference, by positing a universal "humanity" that is prior to the differentiation of people into various cultural, national, or ethnic identities. Recognition of what we have in common is what enables us to tolerate the differences among us.

Postmodern critics charge that such "essentializing" doctrines of "universal humanity" were in fact little more than ideological smokescreens designed to conceal a hidden agenda: to obliterate cultural difference by forcing everyone to conform to the image of white, property-owning, European males, the people who defined "universal" humanity. Hence, the emancipatory program of Enlightenment humanism proved to be a fig leaf to cover the naked predatory intentions of European imperialists and colonists bent upon "civilizing" (i.e., subjugating) dark-skinned natives, whose lands Europeans coveted and whose attitudes and practices posed a threat to the identity of those same Europeans.

In view of such an interpretation of the Enlightenment project, some postmodernists reject the idea of "universal" human traits and instead emphasize "celebrating difference" and promote "cultural diversity." The history of the twentieth century shows quite clearly, however, that when people abandon—as Germans did in the 1930s—the search for what all have in common, there is no celebration of difference; instead, difference is *absolutized.* Without being encouraged to posit that one's long-standing ethnic rivals are, after all, human beings first, and Kurds, Romanians, Jews, Palestinians, Irish Catholics—take your pick—second, the fear and loathing so characteristic of ethnic rivalry threatens to take a violent turn. We are witnessing this process in Eastern Europe today, now that the influence of Marxism's universal conception of humankind has lost its grip.

When deep ecologists encourage wider identification and speak of biocentric egalitarianism, they are encouraging us to recognize what we have in common with the non-human inhabitants of this planet. Without recognition of what we have in common with plants, animals, and the land in general, we tend to abuse nature, to regard it as threatening and alien—so different that it must be dominated so that it can be made into something recognizable: products for enhancing our own human aims.

The presupposition that all beings share something in common is not a specifically Western concept. Mahayana Buddhism affirms that all sentient beings have Buddha nature. Moreover, traditional peoples respect other forms of life because they regard them all as coming from the same higher, sacred source. In my view, just as it is not possible to establish a relationship with something that we experience as wholly foreign or alien to us, it is not possible to establish a relationship with something which we reduce to a simple extension of ourselves. The quest for unity and the search for relationship with the other are not mutually exclusive; they are different aspects of the human longing to be at home in a world that is at times confusing and fearful, as well as awesome and beautiful. Hence, it is appropriate to "think globally" while "acting locally." It does make sense to seek a wider identification with things, while simultaneously remaining open to how they are different from ourselves. These are not easy tasks, but they are required of us at this critical time. Undertaking such tasks is what "the future of ecology" is all about.

NOTES

1. George Sessions, "Ecocentrism, Wilderness, and Global Ecosystem Protection," page 36 of a manuscript to be published in *The Wilderness Condition*, ed. Max Oelschlaeger (San Francisco: Sierra Club Books, 1992), 36.

2. Wendell Berry, *The Unsettling of America: Culture and Agriculture* (New York: Avon Books, 1977); Wendell Berry, *The Gift of Good Land: Essays Cultural and Agricultural* (San Francisco: North Point Press, 1981).

3. Thomas Berry, "The Futility of Global Thinking," *Harper's*, September, 1989, 16-22. Cf. also "Out of Your Car, Off Your Horse," *The Atlantic Monthly*, February, 1991, 61-63.

4. Berry, "The Futility of Global Thinking," 18.

5. Alan Atkisson, "The Utility of Global Thinking," *In Context*, No. 25 (Late Spring, 1990), 55–57. I have benefitted from personal discussion with Alan Atkisson about these issues.

6. *Ibid.*, 56.

7. Letter by Wendell Berry, *In Context*, No. 27 (Winter, 1991), 4.

8. Cf. Alan Atkisson's letter replying to Berry, *Ibid.*

9. Jim Cheney, "The Neo-Stoicism of Radical Environmentalism," *Environmental Ethics*, 11 (Winter, 1989), 293–325.

10. *Ibid.*, 302.

11. Michael E. Zimmerman, "Deep Ecology and Ecofeminism: The Emerging Dialogue," in *Reweaving the Web: The Emergence of Ecofeminism*, ed. Irene Diamond and Gloria Feinman Orenstein (San Francisco: Sierra Club Books, 1990).

*
**

PART FIVE

RELIGION AND CONSERVATION

*
**

Metaphors for Birthing:
Towards a New Creation Story
for the Age of Ecology

Elinor Gadon

WE NEED A new creation story for the postmodern age, one that will support our emerging consciousness of the earth as a living force and the interdependence of all forms of life. The familiar story we know so well from the biblical account in *Genesis*—with its ethos of hierarchy and dominance over all creation—is outdated, no longer serving us well when life as we know it on our planet is threatened with unprecedented ecological disaster. As mythographer Joseph Campbell informed us in his widely acclaimed television series with Bill Moyers, *The Power of Myth*, the creation story we have in the West based on the Bible, a view of the universe that belongs to the first millennium B.C.E., does not accord with our concept of either the universe or human dignity.[1]

Although it has been customary to consider the biblical creation story as "revelation," of a higher order than mythology, the *Genesis* account is the paradigmatic myth of origin for Western culture, our deeply held notion of how it was "in the beginning and will always be." This time-worn creation myth continues to shape our popular worldview. So, while it may have become commonplace among deep ecologists to honor the Earth as the source of all life and the ground of being, all too often as we struggle to implement our new paradigm in social, economic and political realities, we find ourselves smack up against the ongoing resistance of a materialistic society, grounded in an older concept of nature as resource, as inert matter to be controlled, manipulated, and used up for economic gain.

Ecological activists seem to be in ongoing confrontation with those who set environmental policy in government agencies and corporate headquarters, as well as the majority of ordinary citizens whose votes are so crucial in the state and local referendums that seek to protect the environment—men and women of good will who are seemingly blind to the new paradigm. Why is it so hard to change consciousness in the face of mounting ecological crises as we witness the depletion of our protective ozone layer, the poisoning of our lakes and rivers, the disappearance of the rainforests and the myriad species that make their homes there, growing stockpiles of undisposable toxic wastes, and runaway population explosion?

For one reason, while the dominant ethos suggests that we live in a secular society in which church and state are separate, and that our culture is based on the humanistic principles and rationality of the eighteenth-century enlightenment, the Judeo-Christian worldview continues to undergird Western culture. We continually find ourselves mired down in the miasma of time-worn mythology and symbols that define our identity and relationships. At some deep existential level our mythology is more real than scientific truths we claim to live by.

And then we privilege rational thinking over mythology. The scientific-technological bias of our culture predisposes us to equate literalness with truth, denigrating myth as fantasy and fiction, mere imagination. But few of us live fully in such a rational world. There are human experiences on the personal and cultural levels, in many cases the most profound in our lives, that can only be expressed in symbolic form. Like visual symbols and poetic metaphors, mythical thinking is another kind of language which communicates more fully our inner knowing of self and of relationship to the world around us. "Myth is a symbolic ordering which makes clear how the world is present for humans"[2] and provides us with our basic orientation towards life.

The noted historian of religion Mircea Eliade tells us that symbolic thinking is an essential part of human nature, coming before language and discursive reasoning. "Images, symbols and myths respond to a need and fill a function, [bringing] to light the deepest aspects of reality which defy any other means of knowledge. . . . The mind uses images to grasp the ultimate reality of things."[3] The image of the Creator as God the Father, a stern old white-man ruling autocratically from on high, is deeply embedded in modern Western consciousness.

Since the second half of the twentieth century, with the academic approach to the study of the history of religions and a prevailing theology that proclaimed "the death of God," it has been widely held in intellectual circles that we create the gods in our own image, that theogony mirrors life on earth, and that relations among the gods reflect the local culture with the power struggle as its *modus vivendi.* And yet clearly our biblical creation myth has continued to serve as a model for our attitude towards the earth and all its progeny. Western culture is so firmly embedded in the Judeo-Christian creation story with God-the-Father's exemplary act of bringing forth the earth and its creatures out of the void, and his ongoing dominion over his creation, that even in our increasing sensitivity to environmental issues we too seldom reflect on the ultimate source of our penchant for control, domination and abuse of nature. Our creation story has indeed led us right out of the Garden of Eden into what in our own age is increasingly becoming a barren wasteland.

Creation in the Bible was a historical event, inaugurating the notion of time as linear, relentlessly moving onward from the Day of Creation to the Day of Judgment. Hebrew revelation historicized divine activity in the world. Another very different notion of time as cyclical, an ongoing round of life, death and rebirth was held by prehistoric peoples, and continues to be the view of primal peoples as well as those world religions not based on revelation—Hinduism, Buddhism and Taoism. Respect for the process of life, of which we are an integral part, and honoring death as part of life which continually renews itself, was at the core of prehistoric religion. The radical shift from this point of view to the historical notion of time, embedded in the Hebrew creation story, was later incorporated into Christianity and has driven Western culture for almost two thousand years. The patriarchal Creator God of the Bible, separate from his creation, stern ruler and harsh judge, has become the model for authority.

Because of the disobedience of our primordial parents, Adam and Eve, humankind has been banished from the Garden of Eden, the earthly paradise with its green vegetation, flowering plants and flowing streams for all eternity. Men condemned to toil by the sweat of their brow for their food and women to give birth in sorrow and in pain, their earthly life judged by God, the Father. The combined effect of Neo-Platonic philosophical and early Christian thought on our culture led to the denial of our human nature, our rootedness in the natural world, and to the belief that to live in the natural world is to live in perdition.[4]

Linear time moves relentlessly towards death and the Day of Judgment in dramatic contrast to the ever-renewing cyclical time of worldviews that are not based on historic divine revelation. In the Christian orientation of Western culture, human sexuality is linked with death, rather than generativity, the source of the ongoing round of renewal; female sexuality and female bodies are considered evil.

Cosmogonic Myths

CREATION STORIES MODEL how it was in the beginning and serve two functions: (1) They answer those awkward questions children so often ask—who made the world?, how will it end? and (2) they justify the existing social system and account for traditional rules and customs.[5] Primal peoples tell their stories over and over again to affirm their connection to the cosmos, continually reenacting the cosmogonic events of the first creation through ritual, in this way internalizing and identifying with the cosmogonic creation.

The term "cosmogony" comes from the Greek word *kosmos* which refers to the order of the universe or to the universe as order, and *genesis* which means coming into being, a birth. According to historian of religion Charles Long,

> Cosmogonic myths refer to that power or force which centers and gives definiteness to the life of a human community. It is through a creative event that a new world, a definite order is given to the stuff of history, the environment and the

psyche. It is in the cosmogonic myth these elements
are centered and personalized—which is to say
they are re-presented in a new form. Creation
means the modification of reality in terms of a
particular structure. In the cosmogonic myths a
particular form is enunciated and generalized. In
this manner the dominant note of the creative
order pervading the entire order opens up new
possibilities for the emergence of new patterns.[6]

Every culture has its own story. There are various modalities
of creation: *ex nihilo*, out of nothing; from chaos or from the
cosmic egg; through the union of world parents; out of sacrifice
or through emergence from the Mother Earth. Creation is not
only cosmogonic, but also ontogenetic: in other words, it is
invoked whenever any new thing—from the quickening of an
embryo to the dawning of an image—is brought to life.

The great creation accounts such as the Hebrew *Genesis*, the
Babylonian *Emuna Elish* or the Mayan *Popul Vuh*, deal with the
cosmogonic scenario: explicitly, the origin of all things is their
foundation, it is what their existence has been built upon.[7]
Some, like the creation act of the Hebrew God who creates from
nothing, are a symbolic *tour de force* against the impacted
empirical cultural histories, and provide a basis for a new
founding, ordering of the world and the human community.

The power of the deity in myths of this type
establish the cosmos as unrelated to and discon-
tinuous from all other structure prior to the statement of the creation of the cosmos and the human
condition as enunciated in the myth. The deity
exists in the void in himself and by himself; the
autonomous and self-created nature of the deity
appear out of the void or out of nothingness
understood as potent reality.[8]

Creation *ex nihilo* seems to be a particularly male way to go
about it, in contrast to the Native American myth of Old Spider
Woman weaving her web, a matrix/matter/womb out of which
all life emerges, a more natural female act. Cosmology reflects

worldview as well as the other way around. We can historically document the male usurpation of creation in the Babylonian epic of creation, the *Emuna Elish*. The god Marduk brutally murders Tiamat, mother of all beings, reorganizing the world out of her dismembered corpse. Shades of *American Psycho*, a recent sensational novel in which the murderer-hero is preoccupied with the dismemberment of his female victims, a familiar pattern in sex crimes!

Another patriarchal conception of creation is through sacrifice, the pastoralist's myth. There seems to be little evidence for blood sacrifice among prepatriarchal neolithic goddess cultures like Çatal-Hüyük in Anatolia, Old Europe, or the Indus Valley. According to pastoralist's view of creation, in the first sacrifice a man and an ox (or bull) were sacrificed. This couple, primordial man and primordial animal, forms a complete unit of society from which the physical world and the social world were created. As they were dismembered, society came into being—from the man came men; from the ox, animals. Thus the total social world was inaugurated in the first sacrifice. In each successive sacrifice, the pattern started in the myth is repeated, men-and-animal being offered up to produce future men and animals. The Indo-Aryan account of creation in the *Ṛg Veda* names that which is most highly prized by the society—cattle and sons—as the reward for sacrifice, a sacrifice that repeats in ritual the events described in the creation myth.[9]

The creative myth of earth in the beginning

ANOTHER MODEL OF creation focuses on the creative role of the earth in the beginning. Emergence myths describe the earth as containing within itself all of the potencies of life. Native American peoples of the Southwest believe that there was no break with the powers of the earth when humankind emerged. The basic motif of these myths is not how the earth came into being but the symbol of earth as the source of all life and forms.

Perhaps this is how our earliest fully human ancestors in the late Paleolithic understood the origin of all things. While it is problematic to interpret the silent evidence of archeological finds, analysis of the art and artifacts of these people in their cultural context suggests that they understood the earth as the source of all life and ground of being. Hunters and gatherers with their highly developed sense of observation, noted the life process in the women of their community—menstruation, pregnancy, childbirth, and lactation—and by analogy imaged the earth as female, a great womb out of which all life emerged, nurtured by the Earth Mother who provided all they would need to sustain life. They fashioned small votive images out of bone, amber, clay, mammoth ivory—female figures which exaggerated those parts of women's bodies that had to do with procreation and nurture, the misnamed "Venuses," the earliest Earth Mothers. They also observed that in the fall of the year, vegetation withered and died, and fell away into the earth only to grow up again the following spring, and understood the earth as the great womb to which life returned at death for renewal. In their cave sanctuaries, deep within the womb of the Mother, they painted startling panoramas of the animals of their environment in brilliant earth colors—wooly mammoth and bison, mountain goat and musk ox, wild horse and deer, all potent with charged energy, transforming them into vital, living, breathing presences. "Ice Age peoples perceived themselves as one with the animals, not separate species; both were nurtured, like the rocks and the trees, by the life force emanating from the earth itself."[10] Through their ritual art, the shaman-priests were offering thanks to the Earth Mother for the gift of life she had given to them.

While we cannot recover the words of these prehistoric peoples, their images provide the clues to their creation story. Myth, symbol and ritual are but three interlocking modes of transformation—the myth is the story, image/symbol the visual expression and the ritual the patterned activity through which understanding of the origin of things is internalized.

Such a worldview is still current among primal peoples like the Native Americans who were the first caretakers of our own land and watched, uncomprehendingly, the wanton pillage of the homeland by the white-man. As the Cheyenne say, "It is by the earth that we live. Without it we could not exist. It nourishes and supports us. From it grow the fruits that we eat, and the grass that sustains the animals whose flesh we live on; from it come forth, and over its surface run, the waters which we drink. We walk on it and unless it is firm and steadfast we cannot live."[11]

A cautionary note: While it is tempting in our dire need today to reclaim the Great Mother as an abiding bountiful presence throughout the living world who provides all, we must not overlook the destructive side of nature. World mythologies are replete with accounts of the Goddess in her fury, chaos, anger and rage. For example, the mark of Pele's (a Hawaiian goddess) displeasure is the erupting volcano. Earthquakes, like the volcanic eruptions, are testimony to the ongoing process of creation, as the tectonic plates that undergird our continents move and shift, reforming the great land masses. Cyclones and hurricanes continue to wreak havoc with seeming randomness. Perhaps a more balanced approach would be to ground our new creation story in the full acknowledgment of the unpredictability of the earth's powers, respecting chaos as inherent in creation.

The cosmology of the generating Earth

THE GODDESS IS the Earth, she is the womb but she is also the very topography, the landscape and the soil itself.

> In the unified premodern view, the body of the deity was the pattern of the world encompassing its physical totality, geographical and topographical. All sacred places were contained in her body. At set times during the year the community reenacted the creation of the earth. In this way the power of the Goddess inherent in the landscape that is her body was revitalized and the interconnection of nature, the human and divine realized.[12]

Such belief and ritual continue to prevail among primal agriculturalists worldwide. For centuries Russia had a predominantly agrarian culture.

> . . . The land was called the Mother, and her physical features—natural or manmade—were given maternal epithets. Rivers which run through the immense steps are still called "little moth-

ers": *Matushka* Don, *Matushka* Dnieper. Most important of all is *Matushka* Volga, whose sacred nature is celebrated in the first recorded folk song and in popular ballad as "Our Dear Mother" and "Our Natal Mother."[13]

The soil was sacred. The Russian peasants implored "Mother Moist Earth" for aid in their lives. Self-moistened, self-inseminated, she seems to need no mate.[14] The belief in parthenogenesis, the conception of the mother deity functioning as the sole generative principle, giving birth without the participation of a male counterpart, is widespread.

The Zunis of New Mexico pray to Awitelin Tsita, their earth divinity. "May the rain-maker water the Earth Mother that she may be made beautiful to look upon. May the rain-makers water the Earth Mother that she may become fruitful and give to her children and to all the world the fruits of her being that they may have food in abundance."[15]

The Indians of the Peruvian Highlands still schedule their agricultural operations around the menstrual periods of Pachamana, the Earth Goddess. Special restrictions are observed at times Pachamana is "open," for the life of the community and the cosmos depend on her fertility.[16] And in India, the farming communities of coastal Orissa continue to celebrate the Festival of the Menses of the Earth every year at the Summer Solstice, the hottest time of the year, just before the sowing of the seeds and the coming of the cooling, fertilizing monsoon rains.[17]

The association of fertility and abundance with the female is at the very heart of the worship of the Goddess. In the cosmogonic myths of agricultural peoples, women become epiphanies of the sacred power of the earth. They usually play crucial roles in the rites associated with earthly fecundity. For some agricultural peoples the working of the land became homologous with the sexual act—women are symbolically assimilated with the land, and the plow or spade is an emblem of the phallus. The acts of women have worldly significance, for they channel the effects of the earth's ability to bear fruit. They are reservoirs of the fertile powers of the earth.[18]

We need a new cosmology

THESE MYTHS OF ancient and primal peoples stir us deeply as we grope for a way to resacralize our own relation to the earth. But we need our own story as a spiritual guide, one that will once again ground humans in natural process, reconnecting us to the other forms of life on our planet and the ecosystem that is our home, the living earth. We are the product of our own time and place—Americans of the late twentieth century. We cannot go back to ancient ways. Most of us are not primal peoples: we are no longer agriculturalists living out our days in harmony with the rhythms of nature. We are city-folk living in an urban culture.

Joseph Campbell told us that the problem of our mythology is to relate the scientific worldview to the actual living of life. He suggested that we need a bigger vision that includes both scientific thinking and the wisdom of the past so that ordinary people can lead visionary lives.[19]

The symbol of this new mythology is the Earth as seen from space. We are all familiar with the dramatic photographs taken by astronauts from their spaceship on the historic first moon-landing in the summer of 1969, revealing to earthbound humans for the first time the sight of their planetary home bathed in its own atmosphere. The outline of the continents etched against the ocean shores exactly as we had seen them drawn on our schoolroom globes was an awesome revelation of creation, a new icon for our time.

However, despite its widely popular appeal, this image of the Earth as seen from outer space is problematic as the sacred symbol for our time.[20] The photograph has been mass-marketed, reproduced in every conceivable medium, even reduced in scale to a tiny bauble that can hang from a chain like a protective amulet to be worn around the neck or dangle from a rear-view mirror. The problem is that the NASA picture, product of our most advanced technology, and symbol of our heroic journey into outer space, now touted as our new frontier, is a distancing image of the Earth that does not affirm our interconnection with a vital, living, earthly presence, but rather our separation from it. And furthermore, the photograph of the earth from outer space in all its pristine beauty, denies the reality of humankind's despoliation of its riches.

Our changing consciousness

THE PHOTOGRAPHIC IMAGE spawned another vision, that of scientist James Lovelock's hypothesis of the earth as a single living ecosystem.[21] This notion of the Earth as alive is very different from our former, commonly held belief of the Earth as dead matter, and is changing our relationship to the planet. Lovelock named his hypothesis for the ancient Greek Earth Mother, Gaia, in what for him at that time was a convenient metaphor. But his naming struck a deep chord in an emerging consciousness. The deified earth in Greek mythology was Gaia, universal mother of humankind. According to Hesiod she came into being after Chaos and brought forth of herself the sky, the mountains and the sea.

Frank Waters, ethnographer of the Hopi, credits the changing consciousness to the deep unconscious, the earth itself—the substratum of humankind's essential being re-emerging in archetypes of wholeness and unity. "Our emerging view of all creation as imbued with a common life force: the living waters and living earth, mountains, trees, plants, animals and humankind. All born of Mother Earth, all bound together in an unbroken continuity, in an indivisible unity of both biological and spiritual ecology."[22]

Our changing consciousness is reflected in our dreaming. Psychotherapists report that their patients dream differently than they did thirty years ago; the Mother Goddess is appearing in their dreams.[23] Eliade, commenting on the role of depth psychology in the return to symbolic thinking in the late twentieth century, notes that "every historical man and woman carries on, within themselves, a great deal of prehistoric humanity. [It is the] image of the Mother which reveals—and alone can reveal—her functions, at once cosmological, anthropological and psychological."[24]

Metaphors of birthing

THE PLANET EARTH, our home, is dying. We need to reconceptualize creation. A most appropriate beginning would be through metaphors of birthing. Metaphors affect the ways in which we perceive, think and act and are essential to knowledge of any objective phenomenon.[25] The English language is redolent with birthing words and phrases with their provocative associations. Birth is defined as (1) a state of being—beginning, creation, foundation, genesis, outset, origin, source—and (2) as a process of becoming—break out, break forth, blossom, burgeon, bud, come into existence, commence, dawn, flower, generate, harvest, and yield.

Birth is very much on our collective, cultural mind as birthrates rise again. The popular media is full of stories about the baby boom, featuring television anchor-women as new mothers. The latest cover-girl is a nude moviestar eight months pregnant.[26] Significantly, attitudes toward childbirth and birthing practices are changing among the middle-class; people are reclaiming the sacredness of this cosmic event from the sterile, mechanistic hospital procedure. Holistic natural birth is the model at home or in the hospital, with midwife in attendance, the mother supported by the women of her community, friends and family, as well as the father. The euphemism for all this togetherness is bonding, that is forging human connection, ritual in the making. The resacralization of the birth of the human child is not unconnected to the remythologyzing of the earth. The human child is the microcosm, the living earth the macrocosm—the dynamic process, the mystery of life itself.

New myths are not the creation of individuals; traditionally mythmaking was the collective act of a people, slender threads spun out over a long time, perhaps centuries, part of an oral tradition handed down from generation to generation. Perhaps a more viable mode in our time, when we have lost the art of storytelling, is through the visual and performing arts. Artists are prescient, the prophets of our time. Working from their deep unconscious, from vision, trance and dream, they give expression to the shifting currents in the collective cultural consciousness. We can only know what we can image, imagine.

Women artists are creating powerful new icons of birthing as they affirm the sacredness of their bodily experience. Despite the centrality of the birth of Christ in Western religion, there have been virtually no representations of childbirth in Western art. Judy Chicago, determined to resacralize birthing, initiated *The Birth Project*.[27] Under her direction some five hundred women worked together over a five year period creating embroideries, tapestries and quilts celebrating the act of birth. Chicago drew the "cartoon," the underlying image, presenting different aspects of the universal experience—mythical, cross-cultural, biological—and needleworkers executed her designs in wall hangings which she hopes will find permanent homes in birthing centers. The birthing images are accompanied by the testimonials of the women about their own experience of childbirth.

One of the most powerful is *Crowning*, representing that awesome moment when the baby's head first appears in the birth canal. In Chicago's drawing the birthing image is stark, powerful, an abstracted female focused on the emerging new life. Her hair flows out over her head like branches of a tree. Her toes are spread and extended like roots connecting her to the earth. In another, a quilting over airbrushed fabric, titled *Earthbirth*, the Earth Mother gives birth to herself, extending the primal matter in a great wave spouting from her mouth. Chicago comments, "I was looking for a way to use the physical process of giving birth as a metaphor for the birth of the universe and of life itself. . . . I want to make forms that resonate to our substance, are recognizable at a cellular level, and . . . I want to make the creation of the universe an intimately feminine act."[28]

For Cuban-born Ann Mendieta, the making of her earth-body sculpture was a way of asserting her emotional ties to the earth and to her mythic past.

> My art is a way I establish the bonds that unite me to the universe. It is a return to the maternal source. Through my earth/body sculptures I become one with the earth. I become an extension of nature and nature becomes an extension of my body. The obsessive act of reasserting my ties with the earth is really the reactivation of primeval beliefs . . . [in] an omnipresent female force, the after-image of being encompassed within the womb, [and] is a manifestation of my thirst for being.[29]

In her early earth-body works like the "Tree of Life" series, she used the outline of her body to create the image, cutting or demarcating her silhouette into natural materials and environments. Standing against the tree, her naked body bathed in mud, she becomes a living icon. Like the seed of the tree, the woman holds the mystery of life in her body.

Sculptor Cristina Biaggi and architect Mimi Lobell want to recreate the direct experience of the earth as the mother. Together they designed an earth sculpture of the Great Goddess as a temple that can be entered.[30] The outside would be a large mound. The inside would be a negative shape of a woman giving birth. The figure would be squatting in a birth-giving position, her legs spread apart and drawn to her sides, her arms tightly clenched around her knees. Entry would be through a passageway leading from the outside doorway into her vagina. The inside would be painted with red ocher, the color of the nurturing uterine blood.

Biaggi's extensive research on site at the megalithic structures in Malta and the Scottish islands gave her firsthand experience with sacred architecture that metaphorically embodies the mysteries served. Through the construction of the Great Goddess temple she seeks to reanimate the consciousness of architecture as a habitable structure in harmony with the environment. Biaggi is looking for a site to build her Goddess mound that can be oriented like those of megalithic Britain to the movements of the sun and moon, connecting the cosmic forces with the powers of earth.

> Sculpture in the Western world has lost the mystical magical presence that it had during the Neolithic period when a temple or a sculpture was considered to be the body of the deity. In creating my sculpture, I wish to bring back some of this magic and mystery. I want to create a space that inspires mystery; that evokes the dark caves of the Goddess—places of rebirth and revitalized consciousness.[31]

Other sculptors like Vijali Hamilton work directly from the living rock. She has carved her monumental female forms in the Andes of Peru, in the mountains of Mexico, the hills of southern California and on the cliffs overlooking the Pacific where the high-tide line rises to the base of her sculptured forms. The rhythm of her work is slow, as she develops a resonance with the animals, the rocks, the weather and the earth. The sculpture will ultimately be completed by the cyclic growth and erosion of the natural surroundings.

As her creative interaction with the living rock evolved, she came to view the earth as "moving, flowing energy rather than crystallized static forms as we are taught to see and psychologically validate." She writes that "I open the boulders to show an area of space depicted as sky with the original shape. It is seeing the spirit within matter, connecting the space within the form with the whole space of the universe."[32]

Printmaker Judith Anderson's meditation on the earth Mother, *Missa Gaia: This is My Body* is a poignant reminder of her losses. The earth, the grasses, the seas, and the infinite variety of creatures *are* her body, incarnations of her being and creativity. And all return home to her womb in death, dismemberment, extinction. In the undisturbed rhythm of earth, life and death are intertwined and balanced in a vast exchange of lives. But in this etching, the brooding figure of the great mother echoes that of the Virgin Mary in the Pieta, knees wide in the birthing posture, she gently holds in pity, love and anguish, a glorious lifeless body, flesh of her flesh. The great mother, surrounded by and filled with animals, embodies at once both celebration and profound grief and anger.[33]

These are but a few examples of the veritable explosion of creativity among women artists who, empowered by the philosophy of ecofeminism, make the symbolic link between the sacredness of their bodies and the sacredness of the earth.

Conclusion

WHO CAN DOUBT that we are living in a myth of our own making as did the ancient peoples? Like them our imperative for survival may prove at least to be the catalyst for a fundamental shift in worldview. If we can put aside our old cosmology that has led to destruction of our life source and alienation from our rightful place among all other living forms, we may yet be able to embrace another creation story, more appropriate for our time in the twenty-first century, one which promises the enhancement as well as the continuity of all life on earth.

NOTES

1. Joseph Campbell with Bill Moyers, *The Power of Myth* (New York: Doubleday and Co., 1988).

2. Charles Long, *Alpha: The Myths of Creation* (Chico, CA: Scholars Press, 1963), 14.

3. Mircea Eliade, *Images and Symbols: Studies in Religious Symbolism* (New York: Sheed and Ward, 1961), 14.

4. See Max Oelschlaeger, *The Idea of Wilderness: From Prehistory to the Age of Ecology* (New Haven: Yale University Press, 1991), esp. chap. 2.

5. Robert Graves, *Larousse Encyclopedia of Mythology* (London: Batchworth Press, 1959), v.

6. Long, 23.

7. David Maclagan, *Creation Myths: Man's Introduction to the World* (London: Thames and Hudson, 1977), 5.

8. Mircea Eliade, ed. *Encyclopedia of Religion* (New York: Macmillan, 1986), 2:95.

9. Bruce Lincoln, *History of Religions*, 15:1 (August 1975), 144-145.

10. Elinor W. Gadon, *The Once and Future Goddess: A Symbol for Our Time* (San Francisco: Harper and Row, 1989), 5.

11. George Bird Grinnell, "Tenure of Land among the Indians," *American Anthropologist* 9 (1907), 3.

12. Elinor W. Gadon, "Book Summary: The Once and Future Goddess: A Symbol for Our Time," *The Elmwood Newsletter* 6:4, 15.

13. Joanna Hubbs, *Mother Russia: The Feminine Myth in Russian Culture* (Bloomington, Indiana University Press, 1988), xiii.

14. Ibid., xiv.

15. Eliade, 4:534.

16. Ibid., 538.

17. Frederique Marglin, "Women's Blood: Challenging the Discourse of Development," *Towards Decolonizing Knowledge: From Development to Dialogue*, Frederique Apfel Marglin and Stephen Marglin, eds. (forthcoming).

18. Eliade, 4:538.

19. Campbell, 30.

20. Yaakov Jerome Garb, "Perspective or Escape? Ecofeminist Musings on Contemporary Earth Imagery," *Reweaving the World: The Emergence of Ecofeminism*, eds. Irene Diamond and Gloria Feinman Orenstein, (San Francisco, CA: Sierra Club Books, 1990), 264–78.

21. J. E. Lovelock, *Gaia, A New Look at Life on Earth* (London: Oxford Univ. Press, 1979).

22. Frank Waters, *Anasazi: Ancient People of the Rock* (New York: Crown Publishers, 1974), 11.

23. Edward C. Whitmont, *Return of the Goddess* (New York: Crossroads, 1982), x.

24. Eliade, 1961, 14.

25. George Lakoff and Mark Johnson, *Metaphors We Live By* (Chicago: Univ. of Chicago Press, 1980).

26. *Vanity Fair*, August 1991.

27. Judy Chicago, *The Birth Project* (Garden City, N.Y.: Doubleday and Co., 1985).

28. Ibid., 13.

29. Gadon, 1989, 278.

30. Ibid., 344.

31. Ibid., 346.

32. Ibid., 343.

33. Ibid., 343.

*
**

The "New" Christian Ecology

Susan Power Bratton

Introduction

FROM THE PERSPECTIVE of many environmental activists, contemporary Christianity has been, at best, a weak influence in the popular movements that have brought us greater public recognition of worldwide ecological problems and more stringent environmental regulation. Although Christians celebrated Earth Day with everyone else, they remained hidden in the masses of the socially conscious American middle class. Environmental commentators have, in fact, accused Christianity of stimulating disrespect for the non-human portions of the universe, and have suggested historic Christian thought is one of the philosophical roots of the growing "ecocrisis." Christians have academically responded to the criticisms posed by historians such as Lynn White, Jr., who suggested the transcendent God of the ancient Hebrews reduced Western fear of nature and encouraged dominance over it,[1] and Roderick Nash, who has suggested Christians are either indifferent to wilderness or associate it with the demonic.[2] Christians have proved themselves at least no worse than Western philosophers or members of the other world religions in pursuing a viable ecophilosophy or ecotheology, yet, one still gets the nagging feeling, that as powerful a force as some streams of Christianity have been in defending human rights and working for world peace, contemporary Christianity has not made a strong enough statement concerning the fragmented human relationship with other members of the "creation community" and the growing specter of worldwide environmental degradation.

Periodically, even theologians and philosophers need to ask themselves where they are and how they got there. They also need to ask themselves if there is somewhere else they need to be. After Earth Day is as good a time as any to determine if Christian ecotheology is on the right road. The thesis of this paper is that Christian ecotheology does need to take some new directions and abandon some old ones. Christians have been trying to resolve the same theological problems repeatedly, and thereby have ignored other issues and inhibited their own potential for insightful leadership in the field of environmental ethics.

Cosmology and the critics

THE FIRST POTENTIAL problem in current Christian ecotheology is the continuing effort to justify Christianity to its critics. The battle over the Lynn White essay is being repeatedly fought with very little new ground gained. Christian writers (including myself) fluctuate between speaking to the criticisms of historic Christianity and proving that Christian theology *does* have something positive to say about the environment. Although there is an obvious need for continued production of popular and semi-popular books discussing Christian foundations for an environmental ethic, the notion that Christianity first has to prove it is interested in the creation is inhibiting development of more concrete statements concerning Christian responsibility in regard to specific environmental problems.

Lying beneath the struggle to justify a 2000 year old religion is the question of whether Christian cosmology (views on how the universe was created and operates relative to the divine) is outmoded (or flatly incorrect) and inadequate to present environmental tasks. One camp, recognizing that Christian cosmology and the Christian concepts of God and salvation are interdependent, feels repeatedly called to defend orthodox Christianity against potential replacements—be they Darwinian models, Asian religions, Western varieties of pantheism, or rambling earth goddesses. Another major camp continues to question Christian cosmology (as well as Christian anthropology and Christology) and look for new alternatives that might provide a "better" basis for environmental problem solving. In the latter group are the process theologians, the New Agers, and those attempting to combine elements of Eastern religious traditions with Christian concepts.

Ironically, there is very little evidence traditional Christian cosmology is inadequate or needs major revision in order to provide a foundation for a viable Christian environmental ethic. A second problem, therefore, is that Christian ecotheology is spending a great deal of effort trying to fix something that isn't broken. Two lines of argument may be utilized to support the tenet that Christian cosmology does not need a major ecological overhaul. The first is that Christianity already holds creation to be of great worth. In Genesis 1, God not only blesses the animals along with the humans (a blessing repeated when Noah and the animals disembark from the ark) but the Creator declares the work of creation to be very good and very beautiful even before humankind arises from the dust. This is evidence for the inherent value of the non-human portion of the cosmos. The Bible clearly states that the earth is the Lord's and that God continues to interact with creation (and to take a joy in it) whether or not humans are present. There is no evidence Adam and Eve are the spoiled youngest children of a doting father. Further, God sets Eve and Adam, not just to tilling Eden, but to protecting and preserving it. Human service to the earth is strongly implied in passages about the garden in Genesis 2. God's providence and blessing continue to flow quietly through creation, while creation continues to freely praise God.[3] One can argue that divine love (agape) itself imparts value, and therefore the continuing love of God for the creation provides a continuing affirmation of creation's worth.[4]

The second argument against major cosmological revision is that through history there have been a number of Christian individuals and communities who have had exemplary attitudes towards not just the agricultural or human landscape, but also towards wild nature. Among the best examples are:

(1) The desert fathers and mothers, such as St. Antony of Egypt, who practiced subsistence agriculture, pursued a nonconsumptive life style, and valued friendships with desert wildlife. The desert ascetics aided injured animals, and thought hyenas and lions worthy of Christian ministry. The first monks believed that Isaiah's prophecies of a peaceable kingdom would

be literally fulfilled and real lions and wolves would lie down with real lambs. They therefore thought peaceful human relationship with nature was a sign of holiness. The traditions of the desert fathers still influence Eastern Orthodox and Roman Catholic spirituality.[5]

(2) The early Celtic monastics, including Patrick, Kevin, Bridgit, and Columba, who spent time in spiritual practice in forests, protected such diverse creatures as wolves, deer and birds, and wrote nature poetry. The Celts are characterized by their appreciation of diversity in nature, their emphasis on God as Creator, and their willingness to protect wild creatures from humans, including hunters and woodsmen. A great sign of holiness, among the Celts, was never to have molested a bird.[6]

(3) The Franciscans, who had an appreciation for diminutive organisms, such as swallows, bees and worms, and utilized basic conservation practices in wood harvesting and gardening. Franciscans also protected creatures from human disturbance, and Saint Francis treated the non-human portion of creation as brother and sister. It is said that animals had such an appreciation for Francis, they would jump into his arms.[7]

(4) The Benedictines, who held agricultural labor and good land management in high regard and tried to improve medieval farming methods.

(5) The "plain farmers," primarily from Germanic or continental European Anabaptist backgrounds, who consider care for the land a virtue and limit themselves to "medium tech" agriculture.

(6) Christians, such as John Woolman of the Society of Friends, who showed a concern for the working environments of early industrial laborers as well as attempting to abolish slavery. Woolman refused to wear dyed cloth, since he felt the dyes were potentially injurious to the workers who handled them.[8]

(7) Christians, such as Howard Zahniser, a Free Methodist, who framed the Wilderness Bill of 1964, who have felt dedication to public service in the environmental arena was a legitimate form of Christian calling.[9]

These individuals and communities are associated with a diversity of Orthodox, Roman Catholic and Protestant traditions, and they have appeared throughout the entire span of church history, from the time of the church fathers and mothers to the present day. One would hardly distinguish them as "cosmological experimenters" and several of them are notable for their relative disinterest in scholarly theology.

Historic successes

THE EXISTENCE OF the these "environmentally aware" Christians, in fact, provides an opportunity to look for similarities among Christians who have consciously practiced environmental care or protection. Rather than attack abstract theological constructs, we can ask what factors have encouraged Christians towards positive attitudes and actions, and thus find potential new directions for Christian environmental thought.

The first question, of course, is do these Christians have anything theologically in common, particularly in terms of cosmology? Interestingly, the predominant similarities appear to be in terms of their attitudes towards righteous living and responsibilities towards God and the Christian community. All the above listed Christian "environmentalists" held the pursuit of holiness or personal righteousness as a major goal of the Christian life and considered it critical for the Christian community as a whole to be a righteous witness to the surrounding world. All the groups mentioned, including the Society of Friends and the Free Methodists, value simplicity, although they may have different interpretations of how simplicity is best expressed. Several of the groups view the Christian community as an expression of the Kingdom of God on earth, and several, such as the monastics and the plain farmers, have a countercultural orientation, where they "set themselves apart" in order to become an example to others who are living in the ungodly mainstream.

In terms of relationship to the environment, the primary common denominator is the concept that contact with creation—either through agricultural labor or wilderness spiritual practice—is spiritually beneficial, and that work in, with or for creation forwards holiness or righteousness. The monastic groups believe that creation responds in a positive way to a person who is holy or doing God's will. Therefore a loving relationship with nature is a sign of spiritual purity. In cosmological terms, common themes are an emphasis on the continuing action of God in creation, on a day to day basis, and the potential for creation to disclose something of the character of God. We could, in summary, consider the most important variables sociological rather than theological. The relationship between the individual and the community, the Christian community and the "world," and the Christian community and creation are critical. Further, for these Christians, righteous treatment of nature or a sensitivity to environmental problems becomes an integral part of their total Christian ethic, and is tied to other major spiritual concerns such as care of the poor and avoidance of personal material "idols."

The values and attitudes of these exemplary Christians also provide an opportunity to test the validity of some key hypotheses concerning the supposed weaknesses of Christianity in dealing with environmental problems. The first hypothesis is that the concept of God's transcendence (God is separate from and above the creation) undercuts Christian ability to appreciate the material world. Christians would treat the environment better if they thought God were part of nature (pantheism), or if they thought nature had spiritual power (animism).[10] Ironically, some of the most "otherworldly" Christians, the desert monastics, also had a great appreciation for creation as the work of God. Further, the key in actual Christian practice appears to be not whether one considers God transcendent, but whether one expects God's day to day activity to be evident in creation. Those who expect the divine to act repeatedly through the natural world, and expect to see daily evidence of God's providence in the simple, ordinary and natural, have a well-developed respect for the non-human.

A second hypothesis is that Christian emphasis on "dominion" and creation of humans in God's "image" separates Christians from other creatures, and encourages them to see themselves as despotic rulers over creation. A number of Christian groups with exemplary environmental attitudes, of course, did not use English language Bibles, so the Genesis passages may have had slightly different connotations for them. Further, the writings of European intellectuals, such as Francis Bacon, have made the dominion passage more of an issue during and after the Age of Discovery. What characterizes the "environmental"

Christians, both pre- and post-Enlightenment, is their lack of interest in material gain accompanied by a strong emphasis on caring for the creation or for other people. The importance of simplicity among these groups tends to counteract any view of humans as "lords" of creation, or any declarations that humans have the right to make as much economic use of creation as possible. For some of these gentle witnesses, taking more than is needed would in itself be wrong. If humans are made in God's image, then they will only display it when they behave as citizens of the New Kingdom where the lion rests with the lamb.

A third hypothesis is that Christian repression of sexual desires has led to Christian misunderstanding and mistreatment of nature.[11] The number of celibate communities, not just in Christianity, but in other religions, who have preserved or cared for natural areas or wild creatures speaks against this supposition. The celibate, if anything, seek out their fellow creatures for company.

A fourth hypothesis is that Christian philosophical or cosmic dualism encourages depreciation of the environment.[12] Two of the groups listed above could be considered strongly dualistic in their world views. The desert monks led a constant fight against the demonic and were certainly influenced by Neoplatonism. The pre-Christian Celts may have had pantheistic views (or at least believed in shape-changing or shape-shifting), but they also had a strongly dualistic perception of their spiritual universe. Times of the year were characterized by dark and light, and days were designated lucky and unlucky.[13] The ancient Celts thus easily absorbed Christian concepts of good and evil, and, while their saints fed wolves and praised the beauty of the Irish mountains, they also battled lake monsters and displaced the old religion.

The assumption that those who have ascetic values or a strong urge to challenge the demonic automatically hold the creation in low regard has confused our understanding of early Christian attitudes towards nature. Platonism, in some contexts, does treat nature as either disordered or less than perfect. Creation as unworthy or evil is, however, neither an ancient Hebrew idea nor a theme of the New Testament. Dualism may thus either place the material world in a lower position than human rational thought or the ultimate good, or it may consider creation to be very worthy before God while, like humans, it remains subject to evil. Among the desert and Celtic monks, the Hebrew concept of God's ideal and very good creation predominates. Their rejection of their own material desires does not include a rejection of the surrounding natural environment nor of their wildlife neighbors. The will to fight evil, in fact, seems to forward Christian protection of the environment. The "exemplary" Christians have all either held non-human citizens of creation to be subject to evil and suffering and therefore worthy of Christian compassion and ministry, or they have assumed environmental ills, such as the toxic compounds in industrial dyes, are as worthy of their attention as any other source of human (or environmental) distress. The Celts are an exceptionally interesting case because both their pre- and post-Christian world views were dualistic, yet they have produced some of the finest nature writing of the Western world, and they may well have been the first environmental "preservationists" in European tradition.

Not should we care, but how do we care?

A LAST POTENTIAL problem with contemporary Christian ecotheology is its tendency to deal primarily with generalities about the relationships between God, humans and the cosmos, while hesitating to analyze day to day environmental conflicts. Many Christian books and articles now available provide a Biblical justification for environmental care; very few, however, provide detailed discussion of specific ethical issues, such as the production of CO_2 and acid rain by the industrial nations, growing human numbers, the destruction of tropical forests, and a worldwide decline in biodiversity. Christians must do more than decide that it is spiritually correct to care for creation. An effective environmental ethic goes beyond simple advocacy for care of the earth—it specifically identifies undesirable and inappropriate human activities or interactions with the natural world and suggests corrective action. John Woolman's distress over the toxic effects of dyes was not expressed in a general theological tirade against toxic substances, it was expressed by a decision not to wear suspect clothing.

Recently, while working on a project in the ethics of human population regulation, I decided to search the academic theological literature for journal articles and books on the topic. I found relatively little material. There was a scattering of articles, primarily from magazines such as *Christianity Today* and *Christian Century*, and a chapter or two in a couple of Christian volumes on general environmental concerns. The closest I came to an entire scholarly Christian volume on a related topic (other than on the medical ethics side of contraception) was John Boswell's *The Kindness of Strangers: The Abandonment of Children in Western Europe from Late Antiquity to the Renaissance.*[14] There are, however, numerous secular ethics books on the subject. Despite its importance to environmental ethics, family ethics, and the ethics of missions and development, population has received little academic attention in Christian circles. Although John Knox/Westminister Press plans to publish a book of mine on the topic,[15] it seems like a belated statement in light of the overall importance of population issues to environmental concerns.

There are several Christian groups with primarily environmental ministries, including AuSable Institute in Michigan, the Christian Farmers Federation in Canada, and the Ministry for Population Concerns based in California.[16] Yet there is a continuing need to guide the greater Christian community in terms of necessary actions and personal lifestyle changes that are necessary to solve environmental problems. In order to accomplish this, Christian ecotheology must tackle the key issues and be willing to determine which environmental actions and circumstances are right and which are wrong.

After Earth Day

IN SUMMARY, THIS brief review of past Christian environmental successes suggests that contemporary Christian ecotheology is spending too much time arguing with its critics and fretting over cosmology and semantics (and the concerns of academic theologians). An emphasis on Christian lifestyles and spiritual practice has historically been a more productive approach.

Christian environmental theology needs to take even more of an applied environmental direction. The Christian community needs further guidance on major issues, and choosing the best or most righteous actions is not always easy. We must carefully analyze the cultural, economic, and (yes) the religious roots of today's environmental problems, and construct a Christian environmental ethic outlining possible Christian roles and lifestyles which will demonstrate our love for our human neighbors and for our suffering earth.

If, after Earth Day, we want to develop a "new Christian ecology," we would first do well to understand the values of previous Christian "environmentalists." Instead of deciding our Christian roots are inadequate, we should compare our level of commitment to that of St. Antony, St. Kevin or John Woolman. In the end, it isn't our theoretical statements, but our dedication to righteousness that will count.

NOTES

1. Lynn White, Jr., "The historic roots of our ecological crisis," *Science* Vol. 155, 1203–207, 10 March 1967.

2. Roderick Nash, *Wilderness and the American Mind* (New Haven: Yale University Press, 1967).

3. Susan Bratton, "Christian ecotheology and the Old Testament," *Environmental Ethics* Vol. 6, 195–209, 1984.

4. Susan Bratton, "Loving nature—eros or agape," *Environmental Ethics*, in press.

5. Susan Bratton, "The Original desert solitaire: Early Christian monasticism and wilderness," *Environmental Ethics* Vol. 10, 31–53, 1988.

6. Susan Bratton, "Oaks, wolves and love: Celtic monks and northern forests," *Journal of Forest History*, Vol. 33, 4–20, 1989.

7. See the above paper for a list of references.

8. Fredrick B. Tolles, *The Journal of John Woolman and A Plea for the Poor* (Secaucus: The Citadel Press, 1967).

9. Dennis M. Roth, *The Wilderness Movement and the National Forests* (College Station: Intaglio Press, 1988), 6–12.

10. See the Lynn White essay for this and the next hypothesis.

11. See Richard Cartwright Austin, *Beauty of the Lord: Awakening the Senses* (Atlanta: John Knox Press, 1988).

12. Bryan Norton has expressed this theory several times in our conversations.

13. Fredrick Suppe in course materials for "Topics in English and American Literature: The Celtic World," University of Minnesota.

14. (New York: Pantheon Books, 1988)

15. This volume tentatively titled *Six Billion and More: Human Population Regulation and Christian Ethics* is scheduled for publication in spring of 1992 by John Knox/Westminister Press of Louisville, KY.

16. See Wesley Granberg-Michaelson, *Ecology and Life* (Waco, TX: Word Books, 1988), Chapter 5 for a list of Christian environmental organizations. The address of Ministry for Population Concerns is P.O. Box 9955, Glendale, CA 91226.

*
**

Caring for Creation:
Religion in a Time of Ecological Crisis

Max Oelschlaeger

ON THE FACE of it, the claim that religion has a role to play in resolving ecological crisis seems naive if not patently absurd. Virtually no one, other than a few intellectuals, literary critics, or religious zealots, seems to believe that religion has any pragmatic function at all. Conservationists themselves, trailing in the wake of scholars such as Lynn White, Jr., often believe that Judeo-Christianity—its values, metaphysics, and institutions— is the cause of environmental abuse. Religious believers, on the other hand, usually think of their faith as more concerned with other-worldly issues, such as salvation of the soul, than mundane matters such as pollution of the atmosphere or recycling resources. And, almost uniformly, scientific professionals presume that what environmental crisis calls for is more management— better plans and designs for utilizing the earth's resources, closer monitoring of land use and pollution levels, and so on. Believing in the separation of church and state, conservation-minded voters think that they can elect "environmental Presidents" or legislators who will advance the conservation cause; if nothing else, the Department of Interior, and organizations like the Environmental Protection Agency and Department of Energy, will see to conservation. And finally, intellectuals generally (although there are exceptions) dismiss religion as a dead letter: science is truth and religion is naught but superstition.

Appearances, however, are deceiving. Consider the fact that whatever the environmental achievements in the two decades between Earth Days, on the whole conditions are worse. For example, the Erhlichs brought the issue of population to our attention even before Earth Day 1970. In two decades we have added nearly a billion and half people to the earth. One doesn't have to be a rocket scientist to know that unless something is done to defuse "the population bomb" there is little chance of resolving the global ecocrisis. To take another example, in the twenty years between Earth Day I and XX the conditions underlying the so-called global greenhouse—themselves not entirely unrelated to population levels—have steadily deteriorated. Most fundamentally, global CO_2 levels have increased from 324.5 to approximately 355 parts per million between 1970 and 1990. We now read in *Nature* that climate heating is no longer a hypothesis but an ecological exigency of such magnitude that we must act even in the face of uncertain knowledge. To take a final example, between 1970 and 1990 the Gross National Product (GNP) grew by approximately forty percent; over that same span the Index of Sustainable Economic Welfare (ISEW), which includes variables excluded from the national income accounts such as costs of air and water pollution, loss of farmland, and depletion of nonrenewable natural resources, declined by an estimated ten per cent. How many more decades of such "economic prosperity" can the environment sustain?

In short, two decades of almost continuous action, however well intentioned, indicate nothing more clearly than a failure to stem the drift of Western culture toward ecological breakdown. Yet the facts above indicate something more than the worsening of environmental crisis. One likely reason that conditions continue to deteriorate is that, contrary to the expectation of conservationists, voters, environmental professionals and so on, religion does have a positive role to play in conservation. In any case, my thesis is that religion is a *necessary condition* for the resolution of ecological crisis.[1] I am not claiming that religion alone can solve ecological crisis (that religion is a *sufficient condition*). Technology, environmental science, conservationists and other factors serve essential functions. But until the fundamental role of religion is recognized and *acted upon* there is little chance, given the data we have, for success.[2] (The data also countervail the legitimating narrative of modernism, which assumes the inevitable solution of ecocrisis through technological fixes and economic dynamism; at least twenty years of evidence imply that there are no *strictly* technological solutions within the existing sociocultural framework.)[3]

Obviously we cannot discuss the many obstacles within the church, within organized religion itself, that make this assertion problematic. The festooning of religious behavior in America with dogma and orthodoxy that is death-denying rather than life-affirming gives reason for caution. But pessimism, in any case, is a self-fulfilling prophecy. And the reader should be aware that in claiming that religion has a vital role to play in resolving our ecological predicament I am not advocating any one specific religious credo. Indeed, I hope to appeal to a spectrum of religious believers in a way that legitimates "caring for creation" without violating the autonomy of faith.

***

Looking beneath the facade of ecological dysfunction, and our collective efforts to practice conservation since Earth Day 1970, a few have seen something so faint, so vague and ephemeral as to be almost invisible: namely, that the modern world devalues nature. More explicitly, nature has become an economic commodity; its value is established totally within the context of a market economy. The modern mind understands nature as nothing more than matter in motion that through technological and entrepreneurial prowess can be converted into a consumer's cornucopia. Even so-called nature lovers are often little more than consumers of nature, finding in national parks and mountain ski resorts the "authentic experiences" their lives otherwise lack. Outdoor recreation in America is, after all, big business. Our prevailing world view, as a generation of scholars has shown, rests on the assumption that nature is nothing more than raw material to serve the ends of one species: our own. Consequently, we late-moderns are embedded in a final vocabulary that offers ready-made descriptions of nature and of our relations to nature in a way that closes rather than opens discourse about caring for creation.[4]

We have collectively become *Homo oeconomicus*—that is, actors upon the stage of history prepared for us by liberal-democratic, industrial culture. But our identity as Homo oeconomicus—the mass person, the consumer of modern economic society who lives amid unprecedented material splendor and bends the earth to virtually unrestrained human purpose—

came at a price. We ceased being, as Mircea Eliade argues so brilliantly in *The Sacred and Profane*, *Homo religiosus*: the authentically religious person of prehistory who perceived the living world as shot through and through with value. Nature, in short, was sacred, and the primitive, unlike the modern, tried above all else to live in harmony with the sacred cosmos. The economic person of modernity, in contrast to the religious person of prehistory, believes that nature exists only to serve the interests of the individual. In essence, just as mass and energy are the organizing principles of modern physics, Homo oeconomicus is the theoretical construct around which modern economics orbits and our culture, *mutatis mutandis,* is built.

The road from there, that is the deep past of the Paleolithic, to here, that is the modern age, is long. Elsewhere I have presented the details of the ecological transition from hunting-gathering society to agri-culture and ultimately our own modern culture.[5] Here the limits of space preclude a fuller exposition. The essential point is that Judeo-Christianity, above all else, led us onto the stage of history. Unlike the savage mind, the Judeo-Christian mind found meaning in a time beyond the eternal mythical present. Consequently, the perception of nature as sacred—as the source of value—was destroyed.

Scholars such as Paul Shepard (see *Nature and Madness*) and Herbert Schneidau (see *The Sacred Discontent*) have provided us with penetrating expositions of this epochal transition, explanations that reveal far more than the relatively superficial critique of Judeo-Christianity offered by Lynn White, Jr. By the time of biblical authorship, beginning with the book of J, the Neolithic revolution was history, already some 10,000 to 12,000 years underway. As John Passmore notes in *Man's Responsibility for Nature*, the Old Testament did not lead to the agricultural revolution: rather it justifies that ecological transition, pulling a sacred canopy over the tribes of Yahweh to legitimate the lives they were living.[6] The consequence of that sociocultural transformation, from our vantage point a process known as the "agricultural revolution," was a radical shift in perspective. Nature was no longer sacred, since God the creator was above nature and separate from it, but valueless until humanized. Further, God gave dominion to *Man* over the earth.

When we stand back from the so-called environmental crisis, from the immediacy of polluted water and the global greenhouse, and juxtapose it to the Judeo-Christian faith that was fundamental to the creation of the modern world, we discover the untenability of one of its central premises. Namely, that God designed the earth for *Man*. Providence is simply no longer a tenable thesis if one hopes to maintain any kind of faith whatsoever. Of course, such recognition has been long in coming among the faithful. Seemingly only the ecological cognoscenti have accepted the message that the argument from design is untenable. But the reality of ecological crisis has rendered the belief in providence beyond repair. Today there is no *cognitive alternative* to accepting the biological truth that humankind is part of the web of life.

But scientific judgment does not stop at mere observation of nature and society. There are those who have dared look objectively at humankind's religious behavior. Sociologists argue that the human species is biologically underdetermined: in short, the human gene pool alone is incapable of providing for the reality of specifically human life. Enter sociobiology. Culture provides the stability or order, given by genetic information for infrahuman species, necessary for human life: the *meme* is the replicator of cultural transmission.[7] So viewed, however heretical this sounds to the pious, God is a meme: both a living structure around which human life is organized, and a legitimating narrative (god-talk) which organizes and directs life by giving answers to its most basic questions. In consequence, religion can be seen as a co-adapted complex of religious-memes organized around a God-meme. As George Lindbeck argues, religion is plausibly understood as a "cultural and/or linguistic framework or medium that shapes the entirety of life and thought."[8] God-talk thus occupies, any scientific pretense to the contrary, a position of inordinate importance even within a culture such as our own, where the legitimating narratives of modernity (economic, scientific, philosophic) largely obscure the necessity of religion.

Such a perspective assumes relevance for us in present context, for it begins to fill in the outlines of our thesis: that religion is a necessary condition for resolution of ecological crisis. Religion, or more precisely the fabric of religious belief (its symbols, rituals, and texts), offers possibilities to challenge the economic, scientific, and philosophic orthodoxy that nature is nothing more than resource. More specifically, from a sociobiologically informed perspective, the question is whether Judeo-Christian faith in its many varieties (or any post-Judeo-Christian alternative) can meet the contemporaneous challenges to survival. So viewed, the issue facing contemporary Judeo-Christianity is not one of metaphysics but pragmatics. Norman Gottwald's conclusion to his monumental *Tribes of Yahweh* is apropos. He argues that systems of religious practice and belief "claiming to be based on 'biblical faith' will be judged by whether they actually clarify the range and contours of exercisable freedom within the context of the unfolding social process."[9]

In essence, a Judeo-Christian faith that does not address the reality of life in context is a dead letter. More specifically, a so-called blind faith, oblivious to the circumstances of human existence within and dependence upon the web of life, cannot endure—simply because it is a biological impossibility. But every faith tradition *can* in principle reweave its legitimating narrative in light of ecological exigency. Sociobiology, so often misunderstood as deterministic, reducing human freedom to biological program, underscores this premise. As Richard Dawkins argues in *The Selfish Gene*, despite the resistance of religious meme-complexes to change, manifest in the stifling orthodoxy of death-denying religious practice and institutions, we human beings can defy "the selfish memes of our indoctrination. We can even discuss ways of deliberately cultivating and nurturing pure, disinterested altruism . . . [and evolving beyond our late-modern faith that we are Homo oeconomicus]. We, alone on earth, can rebel against the tyranny of the selfish replicators."[10]

Of course change, especially religious innovation, is never easy. The power of the old and established to resist the new and tentative, even when the necessity of change is related to survival, is well documented. Arnold Toynbee's *A Study of History* reminds us that most civilizations collapse not because of alien invaders but rather through internal insufficiencies: they remain fixed in ways that are outmoded by the continuing stream of life in context. Blind faith, then, is precisely that, an "alienating line of tradition which absolutizes and falsely projects the traditional religious models into eternal idols and specters of the mind."[11]

Clearly, then, those who have dared confront the reality of their faith, that is the fact that religious belief is always located in context, also realize that a faith that does not meet the demands of life, the exigencies of living in a real world, is in danger of collapsing, of becoming empty, irrelevant, and inauthentic. Culture is a dynamic rather than static phenomenon. Accordingly, demands for innovation are inevitably placed upon religious belief, upon the sacred canopy itself. Here, then, is a supreme irony, for the religionist has long feared science and philosophy. But it is precisely scholarship, for example sociology, anthropology, and hermeneutics, itself growing out of the second scientific revolution of Darwin and Clausius, that reinvigorates faith, allows us to see religion through a new lens, allows us to reread the great code inter-textually. Such a perspective on religion entails the recognition that we are both conditioned by and conditioning that history of effects which is religion. People of faith need remember that Western culture itself is simply one form among many and, crucially, it remains in process. We late-moderns have not arrived at any final sociocultural destination.

Religion thus assumes a new significance in a time of ecological crisis when we accept the reality that we are biologically underdetermined. The information encoded in our genes is insufficient for human life; culture fills the void, providing "the instructions," as it were, in terms of which we live. Alternatively stated, culture reflects our biological constitution, providing the means for expression of our genetic potentials. Religion itself, or more accurately the rituals, symbols, institutions, and texts through which religion is objectified, has historically functioned to legitimate the social order by providing a sacred canopy for life. As Peter Berger terms it, "a sacred cosmos that will be capable of maintaining itself in the ever-present face of chaos."[12] But religion itself is set in a dynamic context and must therefore renew itself in order to prevent chaos, that is, the collapse of culture.

Granted, as for example William Leiss argues in *The Domination of Nature*, secular culture has seemingly overwhelmed the Judeo-Christian tradition. The Bible today seems little more than something to be tucked under one's arm on the way to Sunday school, and certainly it has no solutions *per se* for pollution buried within. Yet perhaps things are not so simple. David Tracy, among late-modern hermeneuts, has seen deeply into the nature of cultural crisis, and argues that when literate cultures are in crisis the question of gravity is the one of how to read its most fundamental texts. "The once stable author," as he puts it not altogether facetiously, "has been replaced by the unstable reader."[13] Furthermore, as George Lindbeck argues, religious innovation is not only a possibility but a necessity resulting "from the interaction of a cultural-linguistic system with changing situations." People modify their faith, he continues, "because a particular religious interpretive scheme (embodied . . . in religious practice and belief) develops anomalies in its application in new contexts."[14]

Read against the backdrop of a scientifically informed perspective on religion, we find new meaning in biblical criticism, as for example Harold Bloom's contention that "the primal author J . . . constitutes a difference that has made an overwhelming difference, overdetermining all of us—Jew, Christian, Muslim, and secularist."[15] Similarly, Northrop Frye's brilliant arguments in *The Great Code* assume new meaning. Western culture remains fundamentally a biblical culture. The great code yet underpins our cultural tradition to the extent that our lives are incomprehensible without reference to the Bible, for without that linkage there can be no cogent explanation of how Western civilization has come to be.

In sum, my argument is that in order to understand the anomalies of the present we must begin with the past. Religion itself is caught up in the hermeneutic circle, the confluence of texts, the encounter of sociology, biology, and the other sciences with literary criticism and philosophy. The two cultures have met and become one: the reality of natural history undergirds the great code. We late-moderns are bound up in a history of effects where religion has provided the sacred canopy, defined the ultimate meaning of our lives, the legitimating narratives which justify the sociocultural process. In present context we cannot dwell on that history of effects. The point is that regardless of our own personal biography, our own faith, we are caught up in a culture where religion has had momentous consequence. Indeed, it is deeply implicated in the ecological crisis that confronts us. More importantly, it is inextricably involved in any solution to ecological crisis. We late-moderns, deeply embedded in an ecological crisis of now global proportions, are biologically underdetermined and culturally overdetermined and thus impaled on the horns of a dilemma. For there is no human nature to which we might retreat and find safe haven. Simultaneously, the very culture through which we live out our lives is ecologically pathological.

Given the cogency of my thesis, any solution to cultural crisis begins with rereading the classics. One of the merits of the great code, as David Tracy insists, is that the Bible "has functioned with extraordinary flexibility," lending itself time and again to solutions within a tradition for the exigencies of life. And this, after all, is the most fundamental question of religion. As A. N. Whitehead argues, and Berger, Lindbeck, Tracy, *et al* would surely agree, the religious question is whether "the process of the temporal world passes into the formation of other actualities, bound together in an order in which novelty does not mean loss."[16] The Bible, in short, is a classic (it endures) precisely because of its surplus of meaning, which engenders the possibility of its own renewal.[17]

Leaving aside the complicating variable of all the world's religions, the question before us is "Can religion make a difference in the context of our own liberal-democratic state?" And the answer is "Yes, regardless of religious dispensation, for the creation undergirds the very possibility of the meaning of life." Yet the enormous diversity of religious belief appears to complicate this thesis in several ways. For example, "the creation" does not mean the same thing to all people even within the biblical tradition. How are we to read the creation story in Genesis? Clearly, the great code yields more than one interpretation (and even those who claim that the Bible is inerrant recognize alternative if inaccurate readings). Furthermore, there are religions in modern America that fall outside a biblical tradition. Obviously, the present context is unsuited to even a relatively complete analysis of the notion of caring for creation.[18] However we can at least survey a contemporaneous spectrum of faith, from nature and goddess religionists on one end to Christian fundamentalists on the other, in order to see if they coalesce, despite their differences, around a common core of caring for creation. More fundamentally, we can examine the religious middle that increasingly shows sign of affirming the sacredness of life on earth.

Undoubtedly, America's **nature religionists**—for example, Henry Thoreau and John Muir—care for creation. In her masterful study, *Nature Religion in America*, Catherine Albanese documents the history of a persistent American tradition that has understood nature as sacred. Indeed, she concludes that the continual presence of nature religion "is one more sign that, in a 'secular' society, the search for the sacred refuses to go away."[19] Similarly, Mark Sagoff argues in *The Economy of the Earth* that in America, even from the Colonial days, nature was historically a symbol of the divine; the wilderness assured the colonists "of their special relation to God."[20]

Of course, even within nature religion there was and is diversity. For some, such as John Muir, nature was a visible manifestation of God incarnate. For others, such as the New England Transcendentalists, epitomized by Ralph Waldo Emerson, nature itself was not God but was symbolic of the Cosmic Creator, a material emblem for the spirit of God. Interestingly, Albanese finds in nature religion a range of belief from those who sought to live simply and piously in a harmonious relation with nature to those who believed that the harmony of nature itself allowed humankind to dominate the land and its flora and fauna. Some nature religionists, she argues, remained enframed within the outlines of conventional Judeo-Christian orthodoxy, believing that God had made the creation to serve his children, fashioned in his image. Yet nature religionists, such as the contemporary David Douglas, make the obvious counterargument. If we believe that the earth is God's creation then "we would act less recklessly . . . , not only in irreversible ecological affairs, but in quieter relationships with the earth and its creatures day to day."[21]

Beyond nature religion, **Goddess feminism** also seems a natural constituency for caring for creation, although outside the Judeo-Christian mainstream. Elinor Gadon's *The Once and Future Goddess* makes the concern of goddess feminists for life on earth, all the flora and the fauna as well as global ecological processes, abundantly clear. Of course, the goddess feminist argument that the abuse and exploitation of nature that has created ecocrisis is historically linked to the oppression and domination of woman is controversial. And yet a wealth of scholarship, such as Carolyn Merchant's *The Death of Nature*, Gerda Lerner's *The Creation of Patriarchy*, and Susan Griffin's *Woman and Nature*, that is not allied with Goddess worship but complementary, gives cogency to the goddess feminist thesis. Similarly, James Lovelock's arguments in support of the Gaia hypothesis lend credence.[22] Goddess worship does not, Gadon argues, mean going back to pagan religion per se. Rather, "the Goddess has reappeared . . . as a symbol of the healing that is necessary for our survival. . . ." In revering the Goddess we "honor all that lives—women, the earth, its manifold creatures— [and thus there is] no longer need to control, oppress, despoil our planet, to make war."[23]

Also crucial are the arguments of **post-patriarchal Christian feminists.** They share with Goddess feminists a similar analysis of patriarchy, and understand the abuse of nature as largely a consequence of the male-engendered attempt to dominate women and nature. Post-patriarchal Christianity, as represented for example by Rosemary Ruether's *New Woman, New Earth*, and *Beyond Sexism and God-talk*, deconstructs the Christian metaphysics of presence, which created *an ontological divide between spirit*, as epitomized by God (Our Father who made his son made in his image) and Man, *and matter*, as epitomized by Nature and Woman. Through her hermeneutic Ruether reestablishes connections for Christians with the earth and sky, the flora and the fauna while remaining within the biblical tradition. She argues that the "Big Lie" of modernity, that human beings are above nature and therefore naturally dominate it, is giving way to the "Divine Wisdom." The Divine wisdom no longer abstracts God from the cosmos and humankind from the Earth. The Kenosis of the Father thus opens the door for "the Shalom of the Holy; the disclosure of the gracious *Shekinah*; Divine Wisdom; the empowering Matrix; She, in whom we live and move and have our being—She comes; She is here."[24]

Ruether's deconstruction of the sanctified religious rhetoric of tradition and the dualism of spirit and matter serves also to dramatize the question of **Christian fundamentalism.** However promising the possibilities of caring for creation on the religious left, the possibilities on the right immediately appear less favorable. An entire generation of environmentalists, following the lead of Lynn White, Jr., recite in a litany of conservationist faith the charge that the root of ecocrisis lies in the Old Testament where God gave *Man* dominion over the Earth. Yet White's thesis itself has not withstood scholarly criticism even inside the environmentalist movement. And his thesis also is contradicted by religious conservatives themselves. For example, Francis A. Schaeffer's *Pollution and the Death of Man: The Christian View of Ecology*, is a spirited defense of the environmental responsibility incumbent upon evangelical Christians.[25] Nature, he argues, has a value in itself because God made it. *Man* (Schaeffer's term) has a primary relationship to God (since *he* is made in God's image), but nonetheless *man* has a biblically based obligation to respect and care for the creation. "Christians, who should understand the creation principle, have a reason for respecting nature, and when they do, it results in benefits to man. Let us be clear: it is not just a pragmatic attitude; there is a basis for it. We treat it with respect because God made it."[26]

In the final analysis, however, **the religious middle** plays, in the context of liberal-democracy, the crucial part. What of mainstream Protestants and Roman Catholics? Can they, and surely they constitute the majority of all Americans, find grounds within their faith to care for creation?[27] The answer is unequivocally yes. Protestant theologians like John B. Cobb, Jr. argue that life is the central, indeed, the fundamental religious symbol for God. Judeo-Christians who destroy the Earth are also killing God, surely the ultimate apostasy, since "God's life depends on there being some world to include."[28] Whatever God means to Christians, God is necessarily the inclusive whole. Accordingly, all creation must be treated with respect. Catholicism also presents a promising future for environmental care and concern. Pope John Paul II has called for a socially just, religiously inspired, ecological ethic. And among Catholic theologians, such as Matthew Fox, we find strong arguments for an ecological ethic, as in his dramatic symbolic equation of the human exploitation and desecration of nature with the crucifixion and resurrection of Jesus. "Mother Earth is being crucified in our time and is deeply wounded. Like Jesus at Golgotha, she is innocent of any crime, 'like us in every way save sin' (Heb.

4:15). . . . Yet, like Jesus, she rises from her tomb every day."29 In short, centrist Catholics and Protestants agree that there is nothing in the Bible to preclude an ecologically informed yet religious view that nature includes God (although the world is not God).

✳✳✳

In his preface to *Religion and Environmental Crisis*, E. C. Hargrove argues that too much energy has been spent either blaming religion for environmental crisis or arguing for one religion as the right solution. His proposal that we instead try to find "ways for major religions to respond to environmental crisis" is all the more cogent, then, set in the context of Earth Day 1990.30 For the twenty years between Earth Days should disabuse us of any notion that solutions are simply a matter of reason, politics, technics or economics. One reason that amelioration of environmental crisis still eludes us, as Hargrove implies, is that religion has a fundamental role to play in conservation, a function that nothing else can fulfill.

If we view religion generally, and the church more specifically, as part of a larger institutional framework that determines sociocultural outcomes, then religion's role becomes even clearer. For not only can religion help us to see nature as something more than a mere resource to fuel the economy and an environmental sink for our effluents, but the church itself—regardless of denomination—is, or at least can be, an important vehicle of innovation. Generally considered, there are four institutions that exercise influence and shape public opinion in America: the state, the corporation, the university, and the church. Between them the state and the corporation are Leviathans of such unprecedented power that they almost singlehandedly determine the texture of modern life. We live, as many observers suggest, in an era of controlled capitalism or state industrialism. The GNP, the rate of economic growth, and the bottom line are the idols which America worships, and the state and the corporation are the high priests of this faith.

To look toward the state to solve environmental crisis is a folly (although it is necessarily involved in any solution). There is more than a little irony in the fact that the single largest source of pollution in our nation is the federal government. And clearly, environmental legislation—energy conservation, solar energy research, protection of endangered species, clean air, and so on—is inevitably compromised by short-term economic interests. Similarly, to look toward the corporation (even waving the banner of corporate social responsibility) seems incredibly naive. In the present ethical and legal environment, the reality of the bottom line almost inevitably short-circuits any potential the corporation has to redirect American society toward sustainability. And faith in the university (even on the occasion of an academic conference on "Continuing the Conservation Effort") to promote meaningful change is difficult to maintain.

A few unrepentant liberals, such as John Kenneth Galbraith, hold out hope, arguing that since the new industrial state is dependent upon the trained professionals that keep it running, and since the university educates these individuals, the system can therefore be subverted from within. But Galbraithian optimism is difficult to sustain for anyone familiar with the inner workings of the modern university, private or public. Increasingly the university is run as the corporation: with managers who, as they allocate academic funds, respond to the needs of the corporate job market; with professorial entrepreneurs who cater to the market for fundable research (driven largely by the department of defense and corporate R&D); and with oversight groups concerned largely with economic efficiency.

Which leaves us, then, with the church. Admittedly, the church has a legion of institutional liabilities of its own. But the church, at least within the framework of liberal-democracy, has one thing that the state, the corporation, and the university lack. That is, the possibility of expanding our cultural conversation beyond the modern world view, beyond the vocabulary of Homo oeconomicus. If there is any way that our culture can redirect itself in a time of ecological crisis, then it must be through some old-new sense of the sacred canopy: the fabric of belief that gives meaning to life, direction to culture.

Can the profane person of the modern world become again Homo religiosus? In becoming Homo oeconomicus we have endangered the very fabric of creation that sustains human life. As Thomas Berry observes in "Economics as a Religious Issue," ecology, economy and religion are inextricably entwined. Nature is "the life-giving nourishment of our physical, emotional, aesthetic, moral and religious existence." As we disrupt the web of life "the fabric falls apart—the human fabric, especially—in both its religious and economic aspects."[31] Every religious community, however insular and dogmatic, participates in the natural community of life on earth. But the conventionally religious person of modernity has lost sight of the sacredness of creation. Nature has become mere environment, a source of resources to fuel the industrial state.

Yet, as I have tried to show, however briefly, there is reason for hope, promise of renewal across the entire spectrum of contemporaneous religious belief. Despite diversity of religious belief, there is a common ground for caring for creation. Of course, we have not even touched on the question of how renewal might begin within the church, and how, once begun, that process of change might itself influence the political process, or redirect technology, or resolve any of a thousand and one other issues that collectively comprise the conservation question. But those are questions to be discussed elsewhere.[32]

NOTES

1. At least in America; the situation may well be otherwise in Europe.

2. John Dewey's *A Common Faith* makes clear that religion has the potential to influence human affairs.

3. Of course, "experts" argue that technology offers solutions to all problems, including population. Regrettably, such contentions are difficult to sustain in light of the facts: the road to environmental crisis has been paved with the best of technological intentions. I am cautiously optimistic that ecologically sensitive vocabularies reflect a desire to at least look before we technologically leap. See John Firor, *The Changing Atmosphere: A Global Challenge* (New Haven: Yale University Press, 1990), esp. 100ff, for a lucid discussion of the issue of "the technological fix."

4. Neil Evernden's *Natural Alien* is unexcelled in analysis of the modernist vocabulary.

5. See Max Oelschlaeger, *The Idea of Wilderness: From Prehistory to the Age of Ecology* (New Haven: Yale University Press, 1991).

6. The conventional critique of Judeo-Christianity overlooks this most salient fact. Lynn White, in short, details only part of a much larger picture of the history of effects that underlies ecological crisis.

7. Richard Dawkins' discussion of the basic thesis of sociobiology is unexcelled. See especially Chapter 11 in *The Selfish Gene*.

8. George A. Lindbeck, *The Nature of Doctrine: Religion and Theology in a Postliberal Age* (Philadelphia: Westminster Press, 1984), 33.

9. Norman K. Gottwald, *The Tribes of Yahweh: A Sociology of the Religion of Liberated Israel 1250-1050 B.C.E.* (Maryknoll, NY: Orbis Books, 1979), 708.

10. Richard Dawkins, *The Selfish Gene* (New York: Oxford University Press, 1978), 215.

11. Gottwald, 705.

12. Peter L. Berger, *The Sacred Canopy: Elements of a Sociological Theory of Religion* (New York: Doubleday and Company, Inc., 1967), 50.

13. David Tracy, *Plurality and Ambiguity: Hermeneutics, Religion, Hope* (San Francisco: Harper and Row, 1987), 12.

14. Lindbeck, 39.

15. Harold Bloom, *Ruin the Sacred Truths: Poetry and Belief from the Bible to the Present* (Cambridge: Harvard University Press, 1989), 3.

16. Alfred North Whitehead, *Process and Reality: An Essay in Cosmology, Corrected Edition*, David Ray Griffin and Donald W. Sherburne, eds. (New York: The Free Press, 1978), 340.

17. See Loyal Rue, *Amythia: Crisis in the Natural History of Western Culture* and Jay McDaniel, *Of God and Pelicans: A Theology of Reverence for Life* for marvelous examples.

18. An extended book-length version of the argument in this paper is in preparation, provisionally entitled "Religion in a Time of Ecological Crisis."

19. Catherine L. Albanese, *Nature Religion in America: From the Algonkian Indians to the New Age* (Chicago: University of Chicago Press, 1990), 201.

20. Mark Sagoff, *The Economy of the Earth: Philosophy, Law and the Environment* (Cambridge: Cambridge University Press, 1988), 133.

' 21. David Douglas, *Wilderness Sojourn: Notes in the Desert Silence* (San Francisco: Harper and Row, 1987), 30.

22. See J. E. Lovelock, *Gaia: A New Look at Life on Earth* (New York: Oxford University Press, 1989).

23. Elinor W. Gadon, *The Once and Future Goddess: A Symbol for Our Time* (San Francisco: Harper and Row, 1989), 376.

24. Rosemary Radford Ruether, *Sexism and God-talk: Toward a Feminist Theology* (Boston: Beacon Press, 1983), 266.

25. Our purpose in present context is best served by not critiquing Schaeffer from the assumed legitimacy of another position (such as post-patriarchal Christian feminism, which deconstructs Schaeffer's hierarchies), but by noting that on his reading of the Bible, Christians are obligated to care for the creation.

26. Francis A. Schaeffer, *Pollution and the Death of Man: The Christian View of Ecology* (Wheaton, IL: Tyndale House Publishers, 1970), 76.

27. Evidence now exists that religion is rebounding in America, and beyond the ground captured by the evangelicals and fundamentalists.

28. Charles Birch and John B. Cobb, Jr., *The Liberation of Life: From the Cell to the Community* (Cambridge: Cambridge University Press, 1981), 197.

29. Matthew Fox, *The Coming of the Cosmic Christ: The Healing of Mother Earth and the Birth of a Global Renaissance* (San Francisco: Harper and Row, 1988), 145.

30. Eugene C. Hargrove, ed., *Religion and Environmental Crisis* (Athens: University of Georgia Press, 1986), xvii.

31. Thomas Berry, *The Dream of the Earth* (San Francisco: Sierra Club Books, 1988), 79.

32. See "Religion in a Time of Ecological Crisis" for further discussion.

Notes on Contributors

SAMUEL F. ATKINSON, Department of Biology and Institute of Applied Sciences, University of North Texas, is Director of the Center for Remote Sensing and Landuse Analyses.

SUSAN P. BRATTON, Institute of Ecology, University of Georgia, is a wildlife ecologist and the author of many scientific and philosophical essays in addition to *The Original Desert Solitaire: Wilderness and Christianity*. She is especially interested in developing a Christian ecotheology.

JENNY CHEEK, Project Manager for Distribution, Mary Kay Cosmetics, Dallas, Texas, chairs the Corporate Recycling Council of Dallas, a voluntary association of corporate leaders organized to promote—through leadership, example, and education—conservation within for-profit enterprises. She also directs in-house recycling at Mary Kay Cosmetics.

KEN DAUGHERTY, Department of Chemistry, University of North Texas, Denton, Texas, teaches analytical chemistry, conducts research on refuse derived fuels, and is the author of "Particle Size Determination of Fly Ashes and Relationship to Trace Elements."

KEN DICKSON, Director of the Institute of Applied Sciences, University of North Texas, teaches environmental science, serves on the Executive Committee of the Science Advisory Board for the Environmental Protection Agency, and is the author of *Methods and Approaches for Assessing the Impacts of Point and Nonpoint Source Pollution on Water Quality: The Trinity River, A Case Study*.

NEIL EVERNDEN, Faculty of Environmental Studies, York University, North York, Ontario, Canada, is a zoologist. Among his many publications are *The Natural Alien* and *The Social Creation of Nature* (in press). Evernden specializes in environmental thought.

ELINOR GADON, Department of Art, California College of Arts and Crafts, Oakland, California, is an art historian. Her publications include *The Once and Future Goddess: A Symbol for Our Time*. Gadon is especially interested in goddess feminism and environmental issues.

PETE GUNTER, Department of Philosophy and Religion Studies, University of North Texas, teaches American environmental philosophy, is a noted conservation activist in the Southwest, and is the author of *The Big Thicket*.

GENE HARGROVE, Chair, Department of Philosophy and Religion Studies, University of North Texas, teaches environmental ethics, edits the journal *Environmental Ethics*, is the author of *Foundations of Environmental Ethics* and the editor of *Religion and Environmental Crisis*.

DOLORES LACHAPELLE, Way of the Mountain Center, Durango, Colorado, is a deep ecologist and mountaineer. Her publications include *Earth Festivals, Earth Wisdom* and *Sacred Land, Sacred Sex*. LaChapelle is actively involved with wilderness preservation in the San Juan mountains and elsewhere.

CURT MEINE, NRC, National Academy of Science, Washington, DC, is an environmental historian and philosopher. His publications include *Aldo Leopold: His Life and Work*. His present research involves the design of sustainable agricultural technologies.

MICHAEL NIESWIADOMY, Department of Economics, University of North Texas, teaches resource economics, and is the author of several studies on economics and natural resource conservation appearing in *American Journal of Agricultural Economics, Land Economics, Journal of Environmental Economics and Management*, and elsewhere.

MAX OELSCHLAEGER, Department of Philosophy and Religion Studies, University of North Texas, teaches the philosophy of ecology, and is the author of *The Environmental Imperative: A Socioeconomic Perspective*, *The Idea of Wilderness: From Prehistory to the Age of Ecology*, and editor of *The Wilderness Condition: Essays on Environment and Civilization*.

ROBERT C. PAEHLKE, Department of Political Science, Trent University, Peterborough, Ontario, Canada, is a political scientist. His most recent book is *Environmentalism and the Future of Progressive Politics*, now available in paperback. In addition to his writing, Paehlke is involved in a burgeoning environmental studies program at Trent University.

GEORGE SESSIONS, Department of Philosophy, Sierra College, Rocklin, California, is a philosopher. A widely published environmental philosopher, his most recent book is the best-selling *Deep Ecology: Living as if Nature Mattered* (with Bill DeVall). In addition to his writing and teaching Sessions is an activist in green politics.

F. ANDREW SCHOOLMASTER, Department of Geography, University of North Texas, chairs the Department of Geography.

GENE SPITLER recently retired after a career with Chevron Oil Corporation, San Francisco, California. Spitler has written several papers on environmental ethics and the corporation, including those published in *Environmental Ethics*. He has also been actively involved in environmental affairs both within Chevron and in the greater Bay area community.

MICHAEL ZIMMERMAN, Department of Philosophy, Tulane University, New Orleans, Louisiana, chairs the department at Tulane. Zimmerman has written extensively on environmental thought, ecofeminism and deep ecology. His most recent book is *Heidegger's Confrontation with Modernity*, and he is presently working on a radical environmentalism manuscript for Univ. of California-Berkeley Press.

INDEX